AF573703

THEY'RE OFF!

THEY'RE OFF!

THE STORY OF THE FIRST GIRL JUMP JOCKEYS

ANNE ALCOCK

J. A. ALLEN
LONDON & NEW YORK

British Library Cataloguing in Publication Data

Alcock, Anne
They're off!
1. Women jockeys – Great Britain 2. Steeplechasing – Great Britain
798'.45'0924 SF336.A2

ISBN 0 85131 299 3

Published in Great Britain in 1978 by
J. A. Allen & Company Limited,
1, Lower Grosvenor Place, Buckingham Palace Road,
London, SW1W 0EL
and in the United States of America by
Sporting Book Center, Inc.,
Canaan, N.Y. 12029.

Book production Bill Ireson

Printed and bound in Great Britain by
Biddles Limited, Guildford, Surrey

To

TONY

with my love

&

for jump jockeys

of either sex,

everywhere

FOREWORD

To say that the Sex Discrimination Act came as a shock to the British racing world would be a serious understatement. A large majority of the men who make their living in that world are, to say the least of it, conservative by nature and their reaction to the idea of female jockeys ranged from genuine horror to chauvinistic mockery – with a fair amount of ribald humour in between.

Lester Piggott, never a man to use two words when one will do, said simply "their bottoms are the wrong shape" and, as usual, he had a point.

The average jumping jockeys' feelings were however based not so much on anatomy as on a strange mixture of selfishness and chivalry. There were, most of us felt, quite enough things to worry about in a steeplechase already without the added nightmare of galloping all over some unfortunate damsel in distress.

So those first intrepid females to whom the changed law gave a long awaited opportunity were, in a real sense pioneers. If their early explorations had ended in disaster the words "I told you so" would have been heard on every side.

In fact they have scarcely been heard at all. Women are now not only an accepted part of the jumping scene but, far more often than not, a welcome part as well. That, when you consider all the circumstances, adds up to a remarkable triumph for the sex and it is the story of that triumph which Anne Alcock tells in this book.

A keen and fearless horsewoman herself Anne is quite rightly proud of what her sister jockeys have accomplished. She describes both their successes and their failures with sympathy born of shared experience and you would have to be a very piggish male chauvinist indeed not to admire the enthusiasm and courage of which she writes so well.

From the safety of the grandstand I congratulate both her and her subjects and if only I could put the clock back twenty years I would be proud and delighted to ride against them.

John Oaksey.

JOHN OAKSEY

CONTENTS

LIST OF ILLUSTRATIONS

ACKNOWLEDGEMENTS

Thanks are due to all the "subjects" who gave their time and information willingly and enthusiastically – and apologies to those omitted who doubtless deserve mention; thanks also to Lord Oaksey for so kindly writing the Foreword; and to all the men and women who make jumping such a wonderful game.

For permission to reproduce the photographs which appear in this book (relevant Plate numbers are bracketed after each name) I would like to thank: Jim Meads (1, 2, 6, 9, 10, 13, 14, 15, 16, 17 and 18); John Grant (5 and 20); Sport & General Press Agency (4 and 19); Photonews (7); Leslie Lane (11); *Sussex Express & County Herald* (8); Racecourse Technical Services (3). All other photographs are from my own collection or were supplied by friends.

My thanks in particular to Dorothy Laird and Mr. Harry Marshall who answered interminable questions uncomplainingly and without whom it would have been impossible to compile so full a picture; to Mr. Cahir O'Sullivan, Keeper of the Match Book, Ireland and Mr. J. Richardson, secretary of the Injured Jockeys' Fund.

My thanks, too, to a patient husband for his unfailing support and encouragement with both my writing and racing.

CHAPTER ONE

HOW IT BEGAN

SHE urged her mount faster, pigtails flying, jodhpur-clad legs pounding the pedals furiously as the bicycle rolled over the pole, left the ground a few inches and landed askew, jarring every bone in the little girl's body. She lost little momentum, tearing into the next at break-neck speed – a yawning open ditch which her battered bicycle had negotiated successfully many times before in the far end of the over-grown garden.

All the time she kept up a breathless running commentary as she staved off challenges from imaginary rivals, her whip hand coaxing her steed, her gold and black jockey cap, bought at the sweet stall of the local point-to-point, crammed down firmly over her eyes:

"Thunderball is neck and neck; Thunderball is going clear . . . Thunderball has WON!" She collapsed puffing in an elated heap, bicycle wheel spinning as it lay on the ground.

* * *

Twenty years later, pigtails gone, make-believe jockey cap exchanged for real, that girl stood in the paddock at Lingfield Park shivering slightly in the biting February wind, tapping her stick against her polished boot as the runners paraded for the hunter chase. It had all happened so suddenly – at last.

The politicians of the day introduced the Sex Discrimination Act on December 29, 1975. In January 1976 the Jockey Club, for centuries a male fortress, were obliged to issue National Hunt riders' permits to either sex on equal terms.

Within days, the first National Velvet-type dream was coming true, swiftly followed by others. By the end of that season 145 horses had been ridden by women and ten had won.

The Sex Discrimination Act put the Jockey Club in a quandary. Suddenly by law, they had to do something for which they were ill-equipped and ill-prepared – but act they did. Women had always ridden in point-to-point races, the all-amateur meetings run by local hunts under Jockey Club jurisdiction – but how could women be protected.

The hurly-burly, rough-and-tumble cut-throat business of professional jockeys racing for a living is a vastly different affair from amateur point-to-point riders racing for fun.

That women on the point-to-point scene could be totally ruthless, giving as good as they got in the few mixed races allowed as well as in ladies' races, was not really discussed then, though it had been in the past.

In 1962 the Jockey Club had introduced a minimum age of 18 for women in point-to-points to allow their strength more time to develop following two fatal accidents. It also made the chin strap compulsory for women only – because of their extra hair! In 1964 all riders had to wear chin straps. Was it because, by that time, it was becoming fashionable for men to sport long hair too?

Faced with the Sex Discrimination Act in 1976 the Jockey Club did not wish to introduce a measure – women jockeys in jumping – which could damage the Club's reputation. If only a handful of women rode in professional races, or even in the Grand National, they wanted to be fair both to the professionals and to the women.

Women really wanted to enter pukka jumping in a *quiet* way. Several, and their male champions, had pressed for ladies' hunter chases. The late Gay Sheppard, clerk of the course at go-ahead Stratford, had offered to stage the final of the ladies' point-to-point championship there – and how proud he would have been to know that the first NH winner to be ridden by a girl was cheered home on *his* course.

There was a possible loophole, though there would probably have been an outcry had the Jockey Club used it. Clause 44 of the

SDA said competitions could be confined to one sex if "the physical strength, stamina or physique of the average woman puts her at a disadvantage with the average man". Hence, presumably, no female boxers would fight budding Muhammad Alis.

The Jockey Club sought medical and legal advice, under the glare of the Equal Opportunities Commission. True, pound for pound, most women lacked the physical strength of men (though some have great reserves of stamina and no end of moral strength) but another factor loomed large to influence the outcome . . . the noble horse himself is responsible for around 80% plus of the race result, so it is not a straight man-versus-woman physical contest. Anyway, women had shown themselves equal or better than men in show-jumping and the tough world of eventing, both of which sports had evolved in more emancipated times.

On January 19, 1976, the stewards of the Jockey Club announced that they had considered the implications of the Act on racing, particularly relating to women participating in Flat, NH and point-to-points. They would seek further clarification. Consultation with other organisations would take place immediately. A statement would be issued within the next few days.

It came the very next day: a bombshell for the Flat racing girls who worked in stables and who had been nurtured carefully since 1972 by the Jockey Club. They would not now be allowed races of their own.

This was speedily sorted out between the Lady Jockeys' Association, the Minister for Sport, the Equal Opportunities Commission and the Jockey Club. There was adequate provision in the Act for club races.

But it was victory day for the jumping girls; they could apply for amateur or professional licences in the same way as men. These would be granted on the same criteria: the stewards must be satisfied about the rider's competence, and applications must be accompanied by a recommendation from a licensed or permit trainer; and it would qualify women to compete on equal terms with men. No licence would be granted if the stewards were of the opinion that doing so would be likely to endanger other riders or horses.

Seven days later six women stepped inside No 42 Portman Square, London, W.1, trod the deep carpeting, were looked down upon by paintings of famous horses . . . and came face to face with the licensing stewards, Col. Piers Bengough and Major Ewan Cameron, both former amateur riders of some repute.

First to enter was Mrs. Sue Horton (about half of the women who were granted NH riders' permits were married, so few husbands seemed against it). With Sue's record – first winner point-to-pointing at 14 before the minimum age came in, five times national champion lady point-to-point rider, and winner of some 40 flat and six hurdle races in Europe, the interview was virtually a formality. Eight and a half minutes later she emerged waving the first NH riding permit to be issued to a woman, signed by licensing officer, Mr. Harry Marshall. The waiting Press photographers were pleased to see that Sue was a stunning blonde!

Looks had not got her to the top of the point-to-point tree but ability, hard work, and much travelling; there was no big stable or rich daddy behind her. Riding one of the shortest in the sport – she was known as the female Andy Turnell – Sue invariably went the shortest way round and timed her runs beautifully, creating confidence in the horses she rode. Now she would show her prowess to the British public and to the racing world! Sue had planned to start at Wincanton the following Thursday but she was thwarted by frost. She had to wait until Ascot on February 18th when she rode Le Toy in an amateur riders' 2½ mile handicap hurdle. There were 28 runners – 27 of them ridden by men!

To mark the occasion when Ascot's obstacles were first tackled by a lady, the executive presented Sue with a bouquet. This was a far cry from the days of the late Duke of Norfolk, who is reputed to have remarked that only "over my dead body" would jumping come to the Royal course. For *women* to jump there would have been unthinkable!

Sue Horton, better remembered, perhaps, by her maiden name of Aston, had won Italy's feature jumping event three times, but her impact on the English scene was less than it might have been. This was mainly because of injury (she broke her pelvis in a point-to-point that year and followed with a broken thumb). Again, since her marriage to John Horton who farmed near the picturesque Wiltshire village of Castle Combe, she had "settled down" and no longer took rides in far-flung corners of the country as in her champion point-to-point days. The responsibility of looking after a husband, home, and two growing step-sons meant fewer races.

Trainer's wife Muriel Naughton and twin sisters Jane and Diana Thorne were all able riders and had no difficulty in obtaining the next three licences. But the other two applicants that day, Mrs. Marie Tinkler and Mrs. Shelagh French, both highly competent riders, received body blows. They would be required to undergo medical examinations because of their "mature years".

Marie Tinkler, qualified vet, former winner of the Queen Elizabeth show-jumping cup, twice successful in the historic Newmarket Town Plate, winner of numerous point-to-points, wife of trainer Colin and mother of rising jump jockeys Colin junior and Nigel, admitted to the Press that she was "over 35".

The medical examination and issue of a permit was expected to be a formality. Her own doctor in Thirsk, Yorkshire, had pronounced her "100% fit" the previous week. Her riding record spoke for itself.

Marie was 51. A week later she was examined by the Jockey Club's consultant Mr. Frank d'Abreu – and he "failed" her. Ethics prevented his discussing individual cases, but he did say that one of his jobs was to act as a "safety valve". He could act in "the best interests" of an older person who might be fit for any pursuit other than race-riding, and in the interests of the Jockey Club to see that racing was not brought into disrepute should, for instance, an older lady rider suffer a serious fall.

The Press were incredulous. Banner headlines proclaimed: "Sorry, Mum! Jockey Club Doc bars Marie, 51"; "'No' to a Lady"; "Tinklers furious as 'unfit' Marie is refused licence to challenge men"; "Marie denied certificate to ride"; "Stewards stop Marie", and "Jockeys' mother refused licence to ride".

The Tinklers were indeed furious, especially husband Colin who defended Marie vigorously: "She's twice as fit as I am – and stronger, too. She's a tip-top rider and she'd see off more than a few of the male amateurs riding at the moment." He pointed out that Marie rode out three lots every morning, played golf several afternoons a week and managed the household efficiently and energetically.

"It really is unfair," he said. "She rides and schools the horses here, schools them on the racecourse and rides regularly in point-to-points."

Marie herself, choking back disappointment, decided she would not fight the decision: "The doctor told me he didn't approve of people over 50 riding in races and that the consequences could fall upon those responsible for issuing licences." Instead she rode with considerable success on the Flat.

Several journalists noted that the new Sex Discrimination Act would hardly seem a credible means of ushering in equal opportunity after this. So far, women had approached the jumping scene with responsibility and commonsense and the stewards had not had to rule in the case of an inexperienced or incompetent applicant.

On the face of it, it seemed that the Jockey Club was making a stand. The oldest professional NH jockey riding at that time was cheery 45-year-old Joe Guest, while that gallant but oft-patched Spanish amateur, the 59-year-old Duque de Alburquerque did not, at that time, have to have a British medical. His Spanish permit was sufficient. All professionals, amateurs riding in professional races, or those who are 35 years old or over have to pass a medical and produce a medical record book for inspection every time they race in Britain. Such riders have to be examined by the course doctor after a fall and, if an entry is made in red, may not ride again until passed fit. The chief reasoning behind the system is to prevent jockeys with head injuries from riding again too soon, thus risking permanent brain damage.

Marie Tinkler accepted the decision without a fight but Shelagh French certainly would not have done so. Luckily for all concerned she was four years Mrs. Tinkler's junior and got her permit on production of her own doctor's medical certificate. The success with which she rode in the first two years amply proved her ability to those who had not seen her in action in the past – she had no peer in the art of timing on the point-to-point circuit and had amassed over 100 winners since she first rode in 1947.

News of Mrs. French's permit came 24 hours after Mrs. Tinkler's application had been rejected and the sporting Yorkshire family promptly wished Kent's mother of three daughters the best of luck. Now the stage was set . . .

CHAPTER TWO

MURIEL NAUGHTON

It was a typical midwinter day with surprisingly fast ground – a taste of the drought to come. The meeting was at Ayr, the only other scheduled for that Friday, January 30, 1976, was that at Doncaster, and this had been abandoned because of frost. Yet again this Scottish track, blessed by the Gulf Stream, had not succumbed to the fickle British weather.

Eight runners cantered down to the start for the second race of the day, the Spittall Hill amateur riders' handicap chase, 3m 110yds, value to winner £590.10. An ordinary chase . . . made extra-ordinary by the presence in it of a woman rider, Mrs. Muriel Naughton. She was riding her own Ballycasey, a 12-year-old allotted the minimum 10st, reduced to 9st 7lb by Mrs. Naughton claiming the full allowance; as it was her first appearance under rules, there was no question of having ridden ten previous winners which would reduce the allowance.

Muriel Naughton, 28-year-old wife of Yorkshire trainer Mick Naughton and mother of five-year-old Sarah, had wasted no time: it was only three days since her permit had been issued.

Muriel was no women's libber. She just felt strongly that if competent women wanted to "have a go", they should be able to do so. This delightfully unspoilt Shropshire born, Yorkshire based girl, shunned all the fuss and publicity, hating the cameras thrust under her nose. She just wanted to be allowed to get on with what she enjoyed best – race-riding.

Muriel displayed no trace of nerves, showed the first fence to her mount, then coollv kept her eye on the starter's rostrum as she circled with the others, ready to anticipate the off and determined not to be caught napping.

There was a nip in the air but she tingled inside. The crowd responded warmly, and the atmosphere was that of an "occasion".

She jumped off smartly as the tapes went up and the race was on, with Scarfell showing the way. Muriel's capable hands restrained her free-running horse, on whom she had won her first point-to-point. She tucked him in behind Graham Macmillan on Scarfell in the early stages, the fences flicking safely beneath her, the old horse giving her a wonderful feel. He jumped to the front on reaching the stands, the crowd humming with excitement, the cameras rolling. Down the back straight for the last time Ballycasey began making mistakes as he weakened. Muriel stuck to him like glue and the pair passed the post ahead of just one other horse, some 43 lengths behind the winner, Old Stephen, ridden by Hugh Barclay. It was a creditable effort on a lowly horse – and a landmark in the history of British steeplechasing.

Muriel was thankful that she had neither fallen nor caused a pile-up. The stewards had seemed concerned about the whole question of women racing when she went for her interview at Portman Square, and had been relieved to hear that she intended riding her own "safe conveyance" at least for a start.

The fact that the first girls had to go to London to be personally vetted before licences were granted irked her a little. It was not the usual practice with men. Before the influx of women riders it had been, virtually, a piece of cake to get an amateur rider's permit. A supporting letter from a trainer was the basic requirement. But once women were in, the stewards set about tightening the rules. Admittedly they did not rush into this nor submit to outside pressure – within hours of the granting of the first ladies' permits, the Jockeys' Association were up in arms about "the threat to their members' livelihoods".

They demanded that amateurs should have ridden in 75 amateur races (excluding point-to-points) before they were allowed to ride against professionals. In an almost indecently hasty reaction, the Association claimed that their aim was to protect the earning capacity of professionals and their safety. They insisted that a riding fee should be paid for these amateurs, to go into a fund to ensure adequate insurance cover.

Resentment of amateurs by professionals had long been prevalent in racing and the Jockey Club had already taken steps to ensure that a successful amateur did not "pinch" rides from professionals. In the old days they would politely ask such riders either to turn professional or to withdraw discreetly from the scene. Later they ruled that after an amateur had ridden in 75 races against professionals, the owner would have to pay a full riding fee – this does not, of course, go to the amateur, but is swallowed up in the administrative costs of racing. The only exception is where the amateur rides his own horse or one belonging to his wife, parents or children.

The Jockey Club mulled over the new situation and changed the rule in July 1977. From that date, two types of amateur riders' permit came into force – Category A confining holders to amateur races only and Category B which allowed them in all races other than those confined to professionals (such as opportunity races for claiming lads). Both licences included amateur Flat races for the first time.

To obtain a B licence a rider had to show evidence of 15 rides under NH rules (far short of the Jockeys' Association's figure of 75 rides). This could include point-to-point winners, or 20 placed rides in point-to-points (firsts, seconds and thirds only). The applicant had to supply this information from his own records with sufficient detail for it to be substantiated.

Even then, it was stressed, the granting of a B permit would not be automatic, but "dependent on the Licensing Committee being fully satisfied that the applicant has gained sufficient experience to ride against professionals."

For some time before the issue of A and B permits the stewards had, in fact, reserved to themselves the right to impose any restriction they wished upon the holders of permits and licences granted by them (as opposed to overseas holders) if they deemed it necessary.

The introduction of A and B permits was a good idea and one that could justifiably have been adopted many years previously.

There were always a few inept male amateurs around. To do it, it took the advent of women (who, mostly, only applied for permits when already well experienced in riding out for trainers or racing in point-to-points). The system seems as fair a way as possible of protecting professionals from the occasional amateur "clown" who may infiltrate their ranks, and the move was accepted by the Jockeys' Association.

Muriel Naughton brought a refreshing outlook to the NH scene. Though she shrank from publicity, there was less of this than the Flat race girls got when they made their debut four years earlier. But there was a subtle difference. While the glamour was highlighted and "glamourised" still more on the Flat, there was a more sinister side. The Pressmen reporting on women in NH racing were waiting, not so much for the first winner (which came in the very first week) but for the hideous fall or glaring mistake. Luckily for the welfare of women in NH racing, neither occurred within the first two years, though Joan Barrow experienced both sides of the racing coin. A fairytale win was followed by a dislocated pelvis, probably the worst of the injuries suffered on the racecourse in that time. Thankfully there were no smashed-in pretty faces or paralysed backs.

It might have happened on February 21. Less than a month after her first public ride Muriel Naughton aimed for a big race, the four mile Eider Chase at Newcastle.

The atmosphere is always good in NH racing, patronised as it is by warm-hearted, friendly people who brave the elements more for the spectacle than for filling their pockets. This is one of the main differences between jumping and the Flat, where betting is big business.

At Newcastle the crowd was alive to the occasion, with a race of over four miles in prospect. The jockeys made their female colleague (on Ballycasey again) feel at ease.

"Will you give us a lead on your good jumper?" she was asked.

There were 12 runners for this £5,000 handicap and Ballycasey was the 50-1 outsider. Muriel was the sole claiming rider amongst those experienced professionals, carrying only 9st 7lb. Quick Reply with Ridley Lamb carried top weight of 11st 8lb followed by the favourite and subsequent winner, Forest King, ridden by David Munro at 11st.

This time Muriel allowed Ballycasey to adopt the front-running tactics he loved. They went bowling along, horse and rider enjoying every moment. At one stage four fences loomed in line ahead, the first an open ditch.

"This really is racing," thought Muriel as Ballycasey attacked it fearlessly, soared over and galloped on towards the next. The thrill of this moment would send tingles of excitement down her spine for years to come.

Ballycasey stayed in front until the eighth and regained that position from the 13th to 15th. He was still disputing the lead approaching the drop fence, the 17th, with about a mile to go.

He took off all right but failed to get his undercarriage down in time to cope with the drop, sending his jockey spiralling spectacularly amongst the flailing hooves as the rest of the field galloped over.

It was the moment other jockeys dread. They don't want to be brought down themselves and don't want to kick a colleague on the ground. Now, for the first time, some of those jockeys recognised that it was the woman rider tangled up amongst the flying legs. It was a fall which looked worse than it was. Muriel escaped with bruises but for weeks afterwards people would hail her with: "Aren't you the lady who fell on TV?"

So Muriel Naughton was not only the first woman to ride in a chase but the first to fall. She was also to be the first to take out a professional licence.

This was a big step and ultimately resulted in jump racing losing her talents altogether. Her husband, Mick, who rode 16 winners from 1963 until a bad fall curtailed his career in 1967, was pleased with her riding and encouraged her to turn pro. He felt it would be a step forward for his yard in Richmond, North Yorkshire, where Muriel was assistant trainer and general mucker-outer. He had seen her hold her own against professionals, warding off unwelcome attentions in a hurdle race at Teesside with one rider trying to squeeze up on her inside while another was intent on chopping her into the rails.

Muriel could take the rough with the smooth and had a charming personality. Her grit had been proved since she was 17. Her father had died and she was left in charge of a farm and stud near Harrogate. It was then that she first met some of the prejudices a woman faces in a man's world.

She found that fellows in the same profession automatically expected her to fail simply BECAUSE she was a woman. Determination to prove them wrong fired her again when the SDA gave her her chance in racing.

Muriel adored chasing, but acknowledged that it might be a different matter on a bad jumper – an amateur is free to refuse the ride! But even having turned professional, a woman like Muriel would still not have felt obliged to take such rides. She put herself first as wife/mother/assistant trainer and only second as jockey. Her livelihood did not depend on racing.

Sadly her professional career under NH rules never materialised. She took out her licence on October 5, 1976, but relinquished it on November 18, without having had a single ride. She had realised that, as a professional jockey, she would not be allowed to run horses of her own.

She did obtain a professional Flat licence and her exertions were channelled in that direction – riding some of the stable's two-year-olds on their first outings and horses which needed "a bit of looking after."

Women had had their own Flat races for four years but Muriel was now part of an even smaller band of women going it alone as professionals on the Flat against the men. Her reception was distinctly suspicious, bordering on the hostile that first year, though in 1977 she felt less and less of a fish out of water.

The stewards' secretaries, those professional men formerly known as stipendiary stewards who are, in effect, the eyes of the stewards advising them on breaches of regulations, foul riding, and so on, kept an extra sharp watch on her movements.

There was a time when the horse next to her at the start played up badly in the stalls and dwelt when the gates opened. Muriel was away quickly and never saw the horse in the race, but smiled when its jockey cursed the bastard as they rode in afterwards. To her surprise, Muriel found herself called aside by the stewards' secretary.

Said he: "I'd like you to see the camera patrol film of that race, I think you crossed the horse next to you as you came out of the stalls and interfered with it."

Muriel was taken aback, but the man was apologising to her a few moments later when he realised that, head on, it appeared that she had obliterated the horse. In fact it had been reluctant to leave the stalls and that was the only reason it did not appear in view.

Muriel's nastiest moment in a Flat race was at Thirsk. The saddle slipped, shooting so far forward that it was up the horse's neck, leaving Muriel in a precarious position.

"My saddle's gone, let me through to the outside," she called out.

"Let her through," shouted an anxious jockey, glancing across and seeing her predicament.

A gap opened and somehow, Muriel pulled up. There was some way to ride back to the finish so she re-saddled her mount, cantered back towards the capacity crowd and put the horse away. Next day, when she read differing accounts in the papers of what had, supposedly, happened to her, she realised that perhaps she should have left the saddle where it was for all to see.

Said Muriel: "I feel the Press are just waiting for women to do something silly. Lester Piggott comes off at various times like anyone else, but if it's a woman the Press make a meal of it – I think there will always be a certain amount of prejudice."

As for women riding over fences, she said: "It's marvellous that there is now an opening for girls who are good enough; they help bring in the crowds and they have done so well. Anything that adds to the spectacle must be good for racing."

Women like Muriel Naughton help to make it good.

CHAPTER THREE

THE THORNE TWINS

WHEN Warwickshire permit trainer and amateur rider John Thorne saw that Sue Horton was applying for a licence to ride, he egged on his twin daughters, Jane and Diana.

"Go on," he said, "I can't ride them all!" And he sent for application forms right away.

So Jane and Diana made up the quartet who got permits to ride. At 21 they were the youngest.

"We don't want you to do anything silly and would rather not see women in the Grand National yet." . . . the stewards spoke to them like Dutch uncles. The presence of a woman in the Grand National the following year, felt the girls, nearly meant curtains for the cause, but there was little the Jockey Club could do to prevent it – other than they had done already by introducing the two types of riding permit.

Like the majority of lady riders who would follow them, Jane and Diana Thorne mostly rode their own horses in hunter chases – occasionally riding one of their young future hunters in a novice hurdle. Their horses were all home-bred and the twins, who broke them in, rode them daily, schooled and galloped them, knew them better than anyone else.

The question of who would ride what was harder to resolve and caused the occasional family squabble. Jane admitted: "Mostly either of us 'bags' one as a youngster and then it's luck how it turns out."

Once a horse is established with a partner it is seldom swopped. They held such a strong hand with good, well-bred horses that when SDA was introduced Jane had already won 12 point-to-points and Diana eight, including the national Goya ladies final on Ben Ruler at the end-of-season Melton Hunt Club meeting in Leicestershire.

The Chesterton Stud is set in flat countryside and is not unduly impressive to the casual eye. But it has earned a valuable reputation on the NH scene, usually housing four or five stallions. These have included two top NH sires – Spartan General (who died in 1978) and the late Indian Ruler. Another, Ritudyr, runs out with his mares on part of the 350-acre farm which is devoted primarily to beef and horses with about 60 acres of corn.

A 60-acre grass field makes a good gallop and a large cattle barn is an ideal indoor school. This is excellent for breaking youngsters, exercising when there is snow, and for using a horse walker – a gadget suspended from the ceiling from which four horses can be attached and left to walk in circles for any length of time – rather like treading the mill. The Thorne racehorses do a lot of road work to compensate for the lack of hills and are hunted hard; the big Warwickshire ditches teach them to look where they are going and to be bold. Occasionally, in bad weather, the Thornes borrow a sand gallop near Warwick and sometimes take the horses swimming in a special pool. This therapeutic exercise has caught on well in recent years.

Evidence that there is more than stud interest at Chesterton greets the visitor who sees flights of hurdles and event type fences lining the paddock by the drive which leads to the farmhouse buildings and large, open-sided stables. Inside, the mellow brick house is lined with sporting trophies and pictures, making it plain that this is very much a family enterprise.

"Was that field ready for drilling or is it still too wet?" John Thorne asks.

"I didn't ride there this morning", says Jane.

"I'll have to get the tractor going – did you get the spare part, Diana?"

Diana had . . . and had also given a pipe-opener to a horse due to run the next week, given instructions to the girls in the yard, and ridden her British eventing representative, The Kingmaker.

After a lunchtime snack of cheese and biscuits the twins would be backing a couple of two-year-olds. In the spring, Jane and Diana would also help with the stud work.

The summer before SDA the Thornes held a party for the twins' 21st birthdays. Seeking something a bit different, it was decided to hold a rodeo during the evening. They had the ideal setting – a corral at the back of the house used for expectant mares was floodlit and enclosed by posts and rails. Some of the unbroken two-year-olds joined in the spirit of things and ensured a high old fling!

Neither Diana nor Jane can decide whether they prefer racing or eventing. They grew up in a racing atmosphere, Jane rather shy like her mother, Diana plumper-faced than her sister. Their father enjoyed his fair share of success and brother Nigel was making a tremendous impact in the short time before he was tragically killed in a car crash when 19.

The twins acquired their taste for eventing via the Pony Club before they were old enough to point-to-point. Before John Thorne knew it, some of his horses ear-marked for racing were patiently learning dressage skills!

This was sometimes a family bone of contention but, generally, the horses with less racing potential found themselves in the demanding, time-consuming eventing field.

An exception was The Kingmaker, a grey by Warwick and one of the few horses bought by the Thornes. One of Britain's leading dressage exponents, Mrs. Lorna Johnstone, bought him as a four-year-old from Ascot Sales, but he proved too strong for her, with a big buck and shy in him.

Jane had an eventer at this time so The Kingmaker went to Diana with no unduly high expectations. He went on to win the British novice championships, the French open championships at Haras du Pin ("the biggest, toughest course I have ever ridden over"), finished second at Badminton and fourth in the 1977 European championships at Burghley, representing Great Britain as an individual and narrowly failing to get a medal. Despite all this activity the Thorne twins found time for racing as well. As soon as they got their permits to ride under NH Rules in that January of 1976 they looked at the hunter-chases for which their horses were entered. Hunter-chases are amateur races confined to hunters holding current qualification certificates. They are a step up for point-to-pointers.

Air General and Ben Ruler were booked for Stratford on February 7 – the first hunter chase of the season. Jane's mare Indian Diva was to run at Warwick a fortnight later. Diana had broken in Ben Ruler. She had won the ladies' point-to-point championship on him the previous year and had schooled round Stratford more than once, so had complete confidence in him. Air General, to be ridden by John Thorne, was considered the better horse but Ben Ruler was in his heyday and was to receive 11lb from him.

"Ben'll beat you, Dad," Diana grinned before the race.

"There are a few others who'll beat us both," he replied.

Not until she arrived at Stratford did Diana realise how much the Press could make their presence felt. They photographed her as she stepped out into the limelight from the tiny caravan beneath the main stand and swarmed round her in the paddock. But once she was on her way to the start she was on her own and not until the tapes had gone up was she confronted with her next problem.

"The men definitely tried to squeeze me out at the first," Diana said later. She resorted to the outside berth while, up front, her father was making the early running on Air General.

Once or twice he looked round to see his daughter.

"Are you all right?" he called out. But once the race was on in earnest he concentrated on the job in hand.

Diana was glad she had let the others get on with it in their early rush. Now, when it mattered most, her horse was still full of running – and a glance across showed that some of her rivals were struggling.

Her father was still going strongly in the lead and Diana went in pursuit, flat out round the final bend towards the last, just one thought on her mind, even more than winning: "I must beat Dad."

In the stands Mrs. Wendy Thorne stood pale and tense with Jane, strangely silent, beside her, nerves on edge. How much worse it can be to watch a loved one racing than to be out there

Plate 1 Sue Horton, the first woman to be granted a permit to ride under NH rules. Sue retired from racing in 1978, due to injuries suffered point-to-point riding, at which she is a former champion.

Plate 2 The first two women to win steeplechase races under NH rules in Britain were the twin sisters Diana (*left*) and Jane Thorne. In April 1978, Jane finished a close second in the Whitbread Gold Cup at Sandown riding Spartan Missile.

oneself – and worse still when a win looks on the cards!

It was obvious now that the race lay between father and daughter.

Air General was the fractional leader as they touched down over the last, but Diana drove Ben Ruler for all she was worth.

"He's letting her win," exclaimed a few disgruntled punters, muffled by the greater roar of encouragement as the pair passed the post locked together, heads down, legs, bodies straining.

John Thorne looked across and saw, for the first time, that his daughter was beside him. He remained adamant, in spite of criticism, that he did not know who was putting in the strong late challenge. They rode back together in their vertical green and black striped colours as the winner was announced – Ben Ruler by a neck! What a family achievement, in the best sporting tradition of jumping. The occasion was marred only by the boos of those who were convinced that father had let daughter win, giving her the honour and glory of becoming the first woman to win a race under NH rules. The Thornes were too excited to hear or notice any dissent.

Diana's historic feat was rewarded in kind at the next Stratford meeting. The firm of Steele and Dolphin presented her with a diamond and platinum horse-shoe ring as a memento.

Two weeks later it was Jane's turn. Incredibly, also having her first ride, she became the second woman to win a race under NH rules – and poor Jane hated the ensuing publicity even more than Diana had done.

Her win was equally meritorious – a wonderful round on Indian Diva in the Air Wedding hunter-chase at Warwick. She was not favourite, but the mare nearly pulled her arms out and jumped impeccably. It seemed over so quickly. She hit the front at the second last and skated away to score by six lengths and two-and-a-half from Greystoke Pillar and Menaphon.

The twins soon saw the other side of the racing coin. Jane came down next time out when favourite, and consequently more nervous. Both fell at Cheltenham. England's premier chase track was to become an unhappy place for women riders in the first couple of years under rules, with a higher proportion of falls to successes there than anywhere else. Neither girl was injured on this occasion. It was at Worcester that Diana's cheek was badly gashed, leaving her with a permanent reminder that falls are not the only racecourse dangers. This injury was inflicted in the paddock – a kick from another horse as she was helping to saddle their runner.

The Nigel Thorne Memorial Hunter Chase at Stratford in March was especially important to the Thorne family. This was confined to owner-riders or to be ridden by the owner's spouse, son or daughter. John Thorne had won previously and was riding Air General again. Jane was on Indian Diva and that mare finished only sixth, John going on to score. Finetta Welton, the only other woman riding, came down at the first on Kenilworth. It was Finetta who, as Mrs. Belcher, enjoyed a run of point-to-point successes with Zanetta and won the 1978 BMW final on her.

The next March it was Jane's turn to beat her father, fittingly, in Nigel's memorial race. The same horses were again concerned, but this time Indian Diva ran on to score, with Armoured Car dividing the stable-companions. It was a fitting tribute to brother and son, and the day may yet come when father and twin daughters take the first three places in that particular race.

Later in 1977 Spartan Missile, another useful home-bred horse, gave Jane a memorable win at Sandown in a race she nearly missed. The horsebox broke down and as Jane stood forlornly by the roadside near Oxford, she was spotted by amateur Jim Wilson, also on his way to the races. He pulled up, gave her a lift – and it was Jim, on Hinterland, that Jane and Spartan Missile beat at the finish of the three mile five furlongs race, a tremendous performance from a five-year-old. At six, he won the Whisky Haig hunters chase (formerly Liverpool Foxhunters) for John, a great feat as one stirrup leather broke at Bechers. Jane fell with him at Cheltenham, but later, in April 1978, finished a praiseworthy second to Tommy Stack on Strombolus in the Whitbread Gold Cup.

Early in the 1977–78 season Jane and Diana were slowly bringing on two home-bred four-year-olds they had broken in, and rode them in a couple of big novice hurdle fields. Jane was on Fury Boy and Diana – who had also been involved in the European eventing championships at Burghley with The Kingmaker – was on Bossman. They had run them a few times the previous year, ridden the other way round, with Fury Boy placing a couple of times.

When they both ran in a huge field at Sandown, 20-1 reflected their chances but Sandown had, of course, already proved a happy place for Jane and she set about creating a surprise result, notching her fourth success under rules in the process. Chesterton Stud and the Thorne family look certain to go on "producing the goods" for a long time to come . . .

Diana's engagement to Nick Henderson was announced in November 1977 and she plans to stay in racing. Nick expected to set up as a trainer on his own in July 1978 – after a June wedding.

CHAPTER FOUR

HOW TO GET STARTED

NOT every girl has the advantages enjoyed by the Thorne twins; for some it is an uphill struggle to get started in racing. It may be trite to say that the most important requisite for a would-be jockey is to *want* to race, yet it cannot be stressed too strongly. It is distressing to see the frightened face of someone who has been pushed or cajoled into racing: there is a different, haunted look compared with the pallor of normal pre-race nerves, the pallor which disappears amid the thrill of it all as adrenalin takes over and the body is transformed into an eager, if amateur, young jockey.

I cannot remember a time when I did not want to race. Surprising, perhaps, in one born and bred in a town (but with a garden large enough to accommodate a battered bicycle!) and whose parents neither owned nor rode horses. Those annual spring visits to the local point-to-points fired my young mind and my determination to participate, one day, never wavered. Many amateur jockeys have grown up with horses, but riding at a tender age is not a necessity. Many a successful professional began with small stature, an admiration for such as Fred Winter . . . and no riding experience. Once started, dedication is the key, as much for the amateur as the professional. The standard is so high that unless a rider devotes everything to the cause, whether paid or not, there is little chance of doing well. Present amateurs include highly skilled assistant trainers like Nick Henderson who, before he retired from riding in 1978 to start training, had the huge advantages of riding and training horses daily and of working under the guiding hand of Fred Winter.

The man or woman who works in an office, but has a burning ambition to race-ride, will compete on equal terms (perhaps with a mere 7lb weight allowance) against such riders. But problems are there to be overcome. If the young secretary arrives in her office at 9am, sporting a track suit and bicycle clips, or with hay wisps still in her hair (rectified in the nearest loo before facing the tame challenges of a typewriter), her workmates will doubtless have dismissed her as crackers long since, and return to filing and polishing their manicured nails without another glance.

After the deep desire to race-ride, the next most important requisite is fitness. This cannot be glossed over; contrary to the average layman's view, a jockey does not simply sit there, with the horse doing all the work! Fitness is more important than sheer riding ability, especially for the newcomer, as race-riding is in a class of its own compared with any other form of equestrianism. Racing ability will develop naturally in a fit person, but will be stifled totally in the unfit. While the top amateurs, men and women, are very good, we still see unfit amateurs at the other end of the scale. They bump in the saddle, hang on by the horse's mouth, and are of more hindrance than help to their mounts.

Such people are undoubtedly riding their own horses "for fun" but, while it is their prerogative to ride thus – if they do not endanger other horses or riders – they could soon discover how much more fun it is to do the job properly. An unfit jockey is unfair both to his horse and to the betting public. Granted, there is unlikely to be much financial support for an unknown amateur, but if getting fit seems a hardship at the time, this is compensated by the subsequent satisfaction on the racecourse.

The best way to get fit for racing is to ride "work" regularly. Stable work and routine also helps but a person who does not own a horse has to find other means.

Make friends with a local trainer and ask if you can come and ride out. If he or a willing head lad has the time and inclination to help and advise you so much the better. If he bawls his head off, you will still be learning – the tough way. The yard's professionals tend to look down their noses at amateurs, but you can learn much from observation – it is worth every minute of the effort, even if the trainer lives some distance away. Most "first lots" are pretty early, so the 9–5 worker should still get back in time.

You will ride different horses and become accustomed to the "feel" and rhythm of a steady work canter, tuning up the muscles across the small of your back and in your thighs. Don't begin by pulling up your stirrups shorter than you have ever ridden before just because the other lads do. Starting a few holes shorter, you will soon find how stiff you become as previously unused muscles are employed. Gradually shorten your stirrups, a hole at a time. In many ways the regular slow canter is better for you than the pre-race pipe-opener at three-parts speed, as this really makes your muscles work. Don't imagine you are racing fit just because you have been hard hunting all winter. You will not have spent five hours in the saddle crossing country at racing length or pace, so different muscles are called into play.

Hunting will, of course, keep you in general trim and help toward overall fitness. And the thrill of jumping across country will sharpen anticipation for your future racing.

Before long the trainer will put you up on a puller. There is a knack in riding a puller, and it is *not* one of the spheres where a woman's lesser physical strength *necessarily* tells against her. Indeed an eight-stone slip of a girl can show up a bloke who regularly gets "carted". The answer probably lies in hands, and the woman's deftness and sympathy. Many men have the gift, but a ham-fisted lad may get nowhere fast. Riding a puller on exercise can be an unenviable task; it is most important not to ride into the heels of the horse in front – and disastrous if you overtake the whole string! But it is extremely good for your arm muscles and is one of the best ways of preparing them for the test of riding a puller in a race.

There are other ways in which you can get racing fit with a horse at home, even if you have only a hunter. One is to put those hours of slogging, steady roadwork to good use – if the horse is reasonably behaved and you can avoid a main road. Shorten your stirrups and practice your racing crouch position, on your knees, legs just about perpendicular, seat off the saddle, body bent forward, head looking straight and hands held just each side of the neck in front of the wither, reins held in a "single bridge."

Though he doesn't have the "feel" of a racehorse, I have been able to practice problem areas of my race-riding on the family hunter, a willing partner in some of my antics. I only make two rules: not to employ him until he is fully hunting fit himself and not to hit him with the whip. Apart from this the field is wide open: long slow canters are beneficial – and when I include a low hurdle or two on the circuit he soon adopts a greyhound image, revelling in the game as he pounds round, giving me that something extra to "sit against" in his eagerness. He patiently allows me to flourish my whip up and down – he is quite used to a hunting whip cracking at his side knowing it will not touch him – and he joins in doggedly as I perform my interpretation of what is meant to be a masterly driving finish.

One of my weakest racing points is to lean too far forward from the waist over fences. This is fine on a good jumper but I am liable to be "pinged off" when a horse pecks on landing or hits a fence hard. Here the hunter comes in handy again, tackling a drop fence with impunity as his rider shoots her legs forward, slips the reins and leans back – performances which no self-respecting racehorse should be expected to suffer!

Non-equine exercises can be performed almost anywhere or at any time of day. Wrists can be strengthened in readiness for riding a puller by squeezing a rubber ball. The stairs are a ready-made aid. Stand on the ground floor, reach up to a stair above your head and pull up by your arms. You could use a local gymnasium, if there is one, where this and a variety of other exercises can be carried out. One of the best pre-racing exercises is the "Hislop Squat" explained by John Hislop in that invaluable guide to every would-be jockey: "*Steeplechasing*". This book is essential reading for expert guidance on every aspect of riding under NH rules, not only for the beginner but as a refresher, too. The exercise is simplicity itself – to squat as if you were adopting the racing position but to hold it when you start to tire and when its benefits really take effect. From this squat you can also move into the "riding a finish" stance, practising keeping your arms and legs in time, seat just clear of your heels.

You can practise whip maneouvres at home or when walking, making sure you master its use in your left hand as well as your right. Don't be in any hurry to use a whip when you start racing. Until you are proficient you will unbalance your horse and do more harm than good. A slap down the shoulder should suffice.

Without clear, sound wind-pipes you will not be able to ride a good race. This is not much of a problem for those who regularly ride fast work, but it is imperative the less fortunate should find other ways – or they will suffer a searing pain through the chest near the end of their first race, especially if riding in a biting February east wind. I find running is best for the wind, especially uphill. Begin by aiming for one target, then increase the distance; finally time yourself, and then improve on that!

All these physical jerks will seem far less of a hardship if you have a definite race at which to aim. This depends on circumstances. If neither you nor a willing member of your family own a horse and if you are otherwise unconnected with racing, you will have to get in the hard way, probably by ringing round trainers. Be prepared for the majority to turn you down as they will not have heard of you but, with luck, the trainer for whom you have been riding out may be impressed enough (not necessarily by your ability but by your determination) to offer you a ride himself.

Failing this you will have to consider buying a horse and this can be done quite suitably on limited funds, by using some basic common sense. Doncaster or Ascot Sales are possibilities. If you have no training permit you will have to put your purchase with a trainer but if you are prepared to do some of the donkey-work yourself he may offer you favourable terms. This way you will get to know your horse really well. Look, listen, learn – and racing's door will gradually creep open for you.

The next step is to obtain your permit to ride. This will be the Category A, restricting you to amateur races, unless you happen to have been placed 20 times in point-to-points already. In this case you may apply for the full Category B permit. You need a trainer or permit-holder to sign the application form, stating that he considers you suitable material to ride in a race.

Equipment is next on your list. You could go to Moss Bros. and arrive at your first race immaculately turned out with new clothes just begging to be christened. But you will be less conspicuous and more comfortable in well-worn favourites. There is no need for most women riders to buy expensive racing boots when their black leather hunting boots (or even rubber ones if you must) will suffice. Most women jockeys do not suffer from the time-honoured jockeys' problem of weight – they are usually too light! Leather hunting boots will give added protection to your shins (which can be badly bruised by stirrup-leathers pressing against thin leather), and will also help to overcome women's problems of underweight.

Cotton breeches are more comfortable than nylon which are inclined to be slippery and somewhat transparent! I don't recommend stretch breeches unless you happen to be particularly shapely. Again, pants will show through if they are very tight, so choose that garment with thought, too! While on underwear, it is sensible to wear a well-supporting bra especially on a horse inclined to jog in the paddock! Another nicety which would never enter a male jockey's head is to match lipstick to colours – there is no point in wearing cherry-red if your colours are orange! Do wear a back protector (it makes you less sweaty if you wear a vest between it and your skin) and tie a stock quite tightly to protect your neck. Tights are best worn under breeches (men wear them, too) and give your shins added protection with elasticated guards.

One of the most useful aids are thin mesh "dressage" gloves with rubber strips made of the same material as rubber reins on the insides; their non-slip power is marvellous and of great help on a puller or a horse which sweats up badly. If your colours are on the big side, or of silk, put elastic bands round the cuffs so that they do not get in the way, and remember a large, strong elastic band to help keep the cap in place on the skull cap.

Goggles require getting used to so practise wearing them at home. They should not be pulled into place until just before you are under starter's orders or they may steam up. You can get them tinted according to the weather but I prefer plain. They help to protect your eyes from cold wind, mud and sandy ground. If they get muddied over you pull them off your face with one quick move . . . something else to practise at home.

Many jockeys are supplied with their equipment by the racecourse valet, but as yet there are no women's valets. If you find yourself short of something like a breast-girth a message can be sent to the valet in the men's changing room and he is sure to oblige. But he may not, as yet, have added spare hairnets to his stock!

The colours will be brought by the owner and the trainer supplies bridle, Irish martingale and boots or bandages for the horse. But the saddle, weightcloth, and either breastplate or breastgirth (essential to prevent the saddle slipping back) are your responsibility. You will know the weight you are to carry well in advance so you can check it with your saddle at home. It is better not to use your racing saddle on daily exercise, but do have at least one gallop or school in it to make sure you are happy with it. If you are a lightweight it is a good idea to invest in a weighted racing saddle as the weight is evenly distributed and will mean that less lead has to be put into a weightcloth, which can be cumbersome. There is a popular racing belief that live weight is better for the horse than dead weight, but amateurs may not be as quiet in the saddle as professionals. Dead weight can be better than live weight bouncing all over the place. I am biased, having always carried a huge amount of dead weight. Our good old point-to-pointer, Rough Scot, once carried 13st to victory, five stone made up in dead weight with the saddle and two bulging weightcloths! One point about light girls in NH racing is that most can do the minimum 9st 7lb, which can be hard for the men.

A word about nerves. This is such a complex matter, affecting people in totally different ways, so I can offer only a glimpse of the subject.

The man or woman with no "nerves" at all is rare (some say the foolhardy fall into this category) though several may profess never to get pre-race jitters. Some "nerves" are a good thing, for they steel the body for what is to come. This produces the extra adrenalin and clarity of mind. Pre-race nerves usually disappear as soon as you are mounted or, at any rate, cantering down to the start. The feel of the horse beneath you gives you confidence – unless he happens to cart you, in which case, bad luck!

I'll never forget the sinking feeling as a powerful horse gathered momentum downhill to the start of a point-to-point. Instead of pulling myself together my legs turned to jelly (probably the result of two bad recent falls from the same horse) and I became as limp and useless as a wet rag, eventually falling in a heap like a sack of potatoes at the starter's feet. He, hardly surprisingly, enquired: "Has she ever ridden before?" Once back aboard and in the fray, more command returned and, even if I was little more than a passenger, the horse went on to beat the country's best ladies' pointer of the time!

Fences can look enormous when you walk the course (and don't skip that important job). But when you show the first to your horse and you are actually looking *down* on the obstacle, it looks far less awesome.

The changing room reveals much about nerves. A normally quiet jockey is suddenly garrulous, the talkative is strangely quiet, the serious-minded becomes a practical joker, though with few women riding at any one NH meeting, this atmosphere is rather missing for them.

I have neither seen nor found any evidence that women suffer more from nerves than men. Probably the best remedy for pre-race nerves is to keep busy; if you are pre-occupied with saddling your horse it will help, and when you are riding in several races there is no time for "between race" nerves. Many people are loners at this time. Friends or relatives who try to make small talk, or who ask infuriatingly: "Are you feeling nervous?" make you even more edgy.

I got a marvellous moral boost before my first ride in a chase. Standing alone on the stand at Lingfield, when it was still something of a novelty for a woman to be riding, I saw the friendly, craggy face of Fred Winter grinning at me. His eyes twinkled as he wished me luck, doffing his brown trilby: "Remember, keep cool, use your head – and enjoy yourself afterwards!" With such expert advice ringing in my ears, how could I possibly have felt nervous!

The experience of regular race-riding reduces nerves to a minimum. Yet before that important race, that hairy novice chaser, or, for some people, that high anticipation of a winner, butterflies still cavort and turn somersaults in the stomach. It is unwise to eat before a race (you probably won't want to, anyway). I once ate a surfeit of glucose tablets and, to my acute embarrassment, nearly brought them up in the paddock. After that I decided I had quite enough energy without resorting to that treatment!

It cannot be sensible to drink strongly to gain Dutch courage, though a "wee tot" on arrival at the course may not be a bad idea, especially if it is cold or wet! If you find you need a stiff drink or three before you race it is time to hang up your boots, hard though this may be for the professional, whose livelihood it is. The amateur should never reach this stage. He risks his neck and cracks numerous bones "for fun" in the first place!

You can help yourself overcome "nerves" by putting mind over matter, something which may require as much determination as riding into the last in a driving finish – and possibly harder to achieve in cold blood. After a sleepless night before a point-to-point race, I felt awful and moped around the next day, trying to "rest" instead of getting on with things. I went out for the race in a virtual trance and got knocked off in a mid-air collision at the first fence! The state of mind is important, and on subsequent similar occasions I told myself sternly: "A race is only six minutes and you can muster all the strength and faculties in the world for that space of time even if you haven't slept for a week." It does seem to work . . .

CHAPTER FIVE

THE PIONEERS

JUDY GOODHEW did the modern pioneering work for women jockeys in NH racing. Her efforts were much maligned and gained small thanks. Judy, the wife of Charles Goodhew, a Southfleet, Kent, permit-holder and gaming club proprietor, was so angry at the appalling way one of their horses was ridden that she wrote for an amateur rider's permit application form, firmly believing she could do the job better herself. That was in 1971, before there were even ladies Flat races. She signed herself simply, J. Goodhew, but when she filled in the form, she disclosed all, signing herself Mrs. Judy Patricia Goodhew. Under the heading "occupation" she wrote "housewife".

The reply was curt and to the point: "The Jockey Club does not grant women licences to ride."

But *The Sporting Life* had spotted the application and printed a paragraph about it. From that, Judy Goodhew was swept up in a wave of publicity – but few women, Mrs. Goodhew said, came forward to champion her cause. The local switchboard was jammed with telephone calls from the Press; one letter reached her addressed simply: "Judy Goodhew, Jockey, England". Dorothy Laird, later to give such a helping hand to lady riders, offered advice, and just two women wrote supporting her effort – trainer's wife Jocelyn Reavey and Susan Ferguson. Otherwise Judy had to battle on her own.

Her application had not been flippant and she was not publicity seeking. She merely applied for a permit to be granted on merit: she had ridden successfully over hurdles in the Channel Islands since 1966 and she had been fourth in the historic Newmarket

Town Plate, instituted by King Charles in 1665 to be run "on the second Thursday in October for ever". Trainers, grooms and jockeys are ineligible for this long-distance race (for which the prizes include Newmarket sausages) but it was, before 1972, the only race from which women were not barred.

Mrs. Goodhew had a safe horse called Pindon whom she wanted to ride in hurdle races and another who, though a bit of a rogue, went sweetly for her. Never one to mince her words, she insisted she simply wanted fair dos, to be "given a mouthful for lousy riding or praise, if due." She felt strongly that there were many hard-working girls in stable yards who deserved rides. Later, when women had started racing, she felt equally strongly that it was the rich ones who could buy good horses and steal the limelight.

Good almost certainly came from Judy Goodhew's lonely campaign. Next year the Jockey Club announced that it would introduce a number of ladies' races, but only on the Flat.

So, at Kempton Park on May 6, 1972, in a sea of mud and a blaze of glory, 21 girls – Judy Goodhew among them – lined up for the mile and a furlong Goya Stakes. It was won by outsider Scorched Earth ridden by Meriel Tufnell, whose three wins in the 11 ladies' races that year made her the first female Flat champion. Second in that first race was trainer's wife Jennifer Barons on Greater. Trainer Sally Hall was third on Harvest Spider and Sue Aston (now Horton) fourth on Straightshooter.

Judy Goodhew *did* achieve her ambition to ride over hurdles in this country though, between times, she had had a baby and did not ride much. Sadly her career ended somewhat ignominously. She was called before the stewards after her horse had played up badly, being told that "in our opinion, this horse is not a suitable ride for an amateur."

Soon after that first ladies' Flat race in 1972, Dorothy Laird suggested that the girls should form their own Association to look after their interests, and she volunteered to be their honorary secretary.

Dorothy was a lifelong journalist and writer who, at that time, worked in the Racing Information Bureau at the Jockey Club's offices in Portman Square.

With help from Peter Smith, secretary of the Jockeys' Association, they based their articles of association on similar lines. Thus the aims of the Lady Jockeys' Association of Great Britain were agreed: to maintain the highest standards of honour, integrity and courtesy among lady jockeys and to support and protect their status and interests generally; to promote and afford to members of the Association facilities for mutual consideration, assistance, co-operation, negotiation and information in all matters concerning lady jockeys; to support the best interests of horse racing in every way possible; only ladies who have weighed out to take part in a race under Jockey Club rules are eligible to become full members with a vote; those who support the aim of the Lady Jockeys' Association may become non-voting associate members upon election. Meriel Tufnel was the first chairman, taking her role seriously and working as hard to make a success of it as she did of her racing.

As chairman of the LJA and first lady Flat race champion she was called upon to "show the flag" at numerous public functions. This led to her enjoying many rides abroad and becoming European lady rider champion in 1973. She was awarded the M.B.E. but gave up her battle against rising weight and retired from racing, opening a riding stables and livery yard.

Dorothy Laird – in what was to become almost a full time post – was the hon. secretary, Vivien Plenderleith treasurer, Patsy Smiles membership secretary and later first vice-chairman, while the committee consisted of Allison Barclay, Diana Bissill, Joan Calvert, Joy Gibson, Rosemary Lomax, Brooke Sanders and Mrs. Beryl Smith.

Dorothy Laird was 60 and, finding daily travelling increasingly irksome, she retired from the Racing Information Bureau on the day the LJA came into existence. From then on the telephone never stopped ringing at her north London home and, time and again, she found her freelance writing playing second fiddle to the demands of the LJA. Yet she loved it, truly infected by the "bug" and totally involved. Dorothy Laird soon started her invaluable records of ladies' races, extending this to NH racing when women entered that field (far harder to follow because none of the races were confined to women only). Without these records neither the Jockey Club, nor this book, would have such detailed statistics!

Dorothy Laird was born and bred in Scotland, combining an inbred love of racing with a passion for boats. Her grandfather, William Stephenson, was an owner and steward in that country and their racing colours have been in the family for over a hundred years. She had no horse of her own, but hunted with the Lanark and Renfrewshire; her first books were on hunting.

Dorothy got a taste for life in a man's world when she worked her way round the world as an ordinary seaman in the *Penang*, a Finnish barque. A model of the ship graces her study. She was with the British Legation in Sweden and Norway during the Second World War and her love of boats emerged strongly afterwards. Fascinated by dockyard sights and sounds she seized an obvious opening, becoming a reporter on the Clyde.

There she met her husband, London ship-owner John Carr who did not really share her love of horses. Because of his position in shipping, Miss Laird felt compelled to give up her dockyard job and reverted to her interest in horses. She had already developed another deep interest: the Royal Family. She was amazed at the influence Royalty wielded despite lack of apparent power. To her delight the Palace gave her permission to write a book about the everyday work of the Queen. *How the Queen Reigns* was the result, followed later by *A Study of the Queen Mother,* showing the changes in the Royal way of life through her time.

It was while a committee member and programme chairman of the Women's Press Club that Dorothy Laird, quite by chance, found herself in racing. It is a policy of the club to take members on outings designed to show them at first-hand some of the subjects about which they might find themselves writing. A day at the races was arranged. Dorothy had maintained her interest in the sport over the years and took notes of colours as a hobby. Now she met owner and international racing journalist David Hedges who, in turn, introduced her to Lord Howard de Walden. An advertisement appeared in *The Times* calling for someone to make racing more of a day for all the family and, almost before she knew it. Dorothy became features editor for the Racing Information Bureau, a post she held until the formation of the Lady Jockeys' Association.

The LJA invited four important figures to join them as honorary members: Peter Smith, secretary of the Jockey's Association, and those three stalwarts who had done so much to get women's role in racing recognised officially in the first place – Florence Nagle, Louie Dingwall and Norah Wilmot.

It was they who had trodden the thorny path on women's behalf. It now seems almost unbelievable that women were not allowed to hold trainers' licences until 1966 and not until the issue had been decided in a court of law . . . Florence Nagle having taken the Jockey Club there in suffragette style. The Club was really behind the times in those days, taking a totally autocratic view. While the rest of the country's women were becoming increasingly emancipated, a woman who wanted to train had to have a head lad hold the licence.

It was not as if women were new to racing: Norah Wilmot, born in 1889, was assistant to her father, Sir Robert, from 1911 to 1931. Yet from then, after the death of her father, until 1966, her head lad had to hold the licence. This in spite of 20 years as assistant and after she and her sister Kathleen had been the first women to ride work on Newmarket Heath – during the First World War!

Also during the Great War, Louie Dingwall, another popular octogenarian trainer four years Miss Wilmot's junior, was a busy motor-cycle dispatch rider in war-torn France! She trained her first winner (in her head lad's name) in 1932.

Florence Nagle, born in 1894, was over 70 when she won the law case which enabled women to hold licences to train. Fine women indeed, to whom all racing people are indebted.

The Jockey Club now pursued a set programme of gradual expansion of opportunity for girls in Flat racing. By having "ladies only" races, stable girls had equal chances, all the girls being granted amateur permits for ladies' races – at that time there was no question of mixed races. The Jockey Club nurtured the girls and followed a policy which encouraged them to work in racing stables at a time when labour was increasingly hard to find.

Stable girls got a chance to race, and by giving them a weight allowance, the Club provided an incentive for trainers to give them race-rides.

Now came a disastrous blow as it became apparent that the Sex Discrimination Act would no longer allow this "discrimination". In future, girls would have to take their chance against men on equal terms by merit. As men had far greater experience than girls could possibly have acquired in a few years of limited races, trainers would obviously prefer to put up men.

Worse still, after having encouraged girls into racing by dangling the carrot of possible race-rides before their eyes, the Jockey Club was suddenly obliged by law to withdraw its programme. Stable girls would be deemed professionals, like lads. No-one in authority had expected the SDA to have this effect, actually *against* women, but the letter of the law, as laid down, was clear.

The LJA sprang into action. As women were heralded into NH racing, bitter disillusionment reigned over the Flat girls' plight. The Jockey Club agreed to see if any way could be found round the new law; the Equal Opportunities Commission was called in and Minister of Sport Dennis Howell agreed to meet the parties concerned.

A deputation met in the Minister's office. In attendance were Dennis Howell, Lord Leverhulme, Charles Weatherby (Jockey Club secretary), Diana Bissill – then LJA chairman, and Vivien Kaye, then secretary.

Several decisions were taken afterwards. It was agreed to change the existing professional/amateur rule so that a professional (i.e. someone who worked with horses) could, so long as he had never held a professional jockey's licence, revert to amateur status after a gap of only a year, instead of three, out of stable employ. This gave the stable girls the chance, if they wished, of relinquishing their jobs to regain amateur status after a year.

But it was not as easy as that. For many, working in stables was their whole life with the occasional race-ride as an additional perk. If they gave up they, probably, could not afford horses of their own to ride as amateurs and would doubtless be unhappy in non-horsey employment. And racing would lose the very labour force it had nurtured.

To help these girls to a decision, the Jockey Club allowed them to continue to ride in races confined to LJA members until 1978. After that, such girls had to ride as professionals or not at all. As club races already existed for such as the Royal Artillery and Fergentri (the international male amateur riders' association), it was agreed that the LJA, as a properly recognised club, could hold races confined to members of its sex. This provided the outlet needed for stable girls, though the male Amateur Riders' Association was against it. Meanwhile, the previously all-male amateur Flat races had to become open to either sex, club races again being excepted.

Women soon made their mark in these races and in 1977 won about two-thirds of them. The performance of one woman, trainer Stan Mellor's wife, Elaine, brought well-deserved praise from the most chauvinistic of male reporters. Modest, but with a professional approach, Mrs. Mellor won the amateur riders' title with eight winners, plus two in Sweden. She had also won the ladies' title in 1975, before there were mixed amateur Flat races.

Since 1978 stable girls have had to make their own way in the professional world. Their path is tough as they still find prejudice at nearly every corner. Few of the lads who set out as aspiring Lester Piggotts ever make the grade as jockeys. If a girl has the added disadvantage of prejudice, then fewer still will reach the top flight. What girls really need is for trainers to support them and give them a fair chance, as Irish trainer Seamus McGrath has done with Welsh-born Joanna Morgan. As a result of McGrath's confidence in her and the opportunities he has given her, Joanna has become third jockey to that powerful stable and has ridden in the Irish Derby. It could be done over here for girls who show flair.

CHAPTER SIX

NO ROOM FOR THE GIRLS

AFTER the Sex Discrimination Act became law the Lady Jockeys' Association quickly opened its doors to the jumping girls, realising that this would give the Association increased strength and bargaining power, especially in view of the imminent loss of some of the Flat girls. Mrs. Nicky Lay was invited to join the committee as jumping representative. The jumping girls demanded little from the LJA. Their biggest bleat concerned changing rooms – or rather, the lack of facilities for females – but they freely acknowledged that racecourses were totally and forgiveably unprepared for the advent of women riders on the NH scene. Had there been a programme mapped out, with ladies hunter chases planned to begin in a certain year on a named number of courses, some plans could have been laid.

So women riders inevitably found themselves in a number of amusing or off-putting predicaments. When Nicky Ledger and I asked the way to the changing room at Lingfield we were escorted to the First Aid Room where we tripped over stretchers and laid out our clothes on blankets (in the fervent hope that these would not be needed) while the official delegated to look after us stood "guard" outside! He proudly escorted us back across the lawn, out of the members turnstile to the weighing room and paddock. After the race he was there again to look after us.

Nicky Ledger had shown ingenuity in obtaining her rider's permit. When she applied as Miss Batchelor, the stewards were somewhat concerned regarding her amateur status.

They feared that she might not be able to afford the fare if offered a ride at the other end of the country and they were unsure whether she was employed as a groom by Roger Ledger. At that time Nicky had been living with Mr. Ledger on his Kent fruit farm for a couple of years, a somewhat delicate situation to explain to the top men in the corridors of power. It was Leap Year and as Miss Batchelor and Mr. Ledger returned home from her abortive trip to Portman Square, Nicky "popped the question" and they were married in Gretna Green style a week later! Nicky rang the Jockey Club with her good news the same afternoon only to be told that proof would be needed. Roger had an Instamatic camera, so he photographed the marriage certificate and sent it by first-class post. The new Mrs. Ledger was riding in her first race almost at once, without having had a honeymoon! Nicky, a girl who always had a passion for speed, was exhilarated by hurdling and felt that women's presence in NH racing was long overdue. She also found that if she gave as good as she got, she soon became accepted by the professionals – Ron Atkins and Bob Champion being two who were full of fun.

Nicky is one of the most regular riders and her perseverance was rewarded in November 1977. She got Bujoji home first at Worcester and a week later the pair were in the line-up for Sandown's £10,000 Mecca Handicap Hurdle.

Changing facilities remained the biggest headache. Muriel Naughton once had to change in the public lavatories which was inexcusable. Nicky Lay, one of the best of the lady riders, found herself at Hereford one day with three rides in quick succession and a "changing room" (the medical room again) too far away for her to cope with the three sets of colours in time. Luckily she knew course manager John Williams, gave him a friendly smile, and emerged changed for each race from his office within the weighing room block – while Mr. Williams discreetly refrained from using it!

A caravan for lady jockeys is probably the best answer and one which has been adopted by several courses. But basic essentials, such as water and loo, are all too often lacking. Ludlow had arranged a caravan for the use of Mrs. Lay and Jane Price, only to find the night before that the caravan had been sold and the new owner was removing it forthwith. Faced with this dilemma clerk of the course Major J. C. G. Moon appealed to that reliable institution, the good old British bobby, asking if the two women might change in the police post. But the heavens opened and the policemen were none too pleased at being turfed out into the rain while their quarters were taken over. Nor was it pleasant for the two women. Wanting nothing more than peace and privacy, they found this disturbed by repeated knocks on the door.

One hot day at Stratford with no valet available for women riders and my boot jack forgotten, I found myself grappling unsuccessfully with a pair of tight boots. The caravan on the concrete beneath the stand was like an oven and I wanted to stay in it as briefly as possible. Looking for help I hailed a patrolling policeman, a burly sergeant with curly moustache. Somewhat taken aback at my strange request, he nevertheless complied, glancing around in embarrassment as he bent over and pulled at my stubborn boots while I sat on the steps.

A valet for the girls could be a good idea (perhaps one of the existing valets' wives could take it on). Though most of the female jockeys have all the equipment they need, there are times when an item is mislaid, a girth proves too long or too short, and a valet could come to the rescue. She could also keep an eye on valuables while a girl is out on the course, and help with the little things like tying the band around the cap, cheering them up after a fall, re-assuring the jittery and generally looking after them. Hot water and trial scales should be provided – but seldom are – and the men usually have tea and sandwiches available for them after racing. They deserve to be pampered, for without jockeys there would be no racing and the changing room is their sanctuary.

The valet tidies up clothes, tucks in back protectors, ties knots in caps, tugs at boots, and can give encouragement to a jockey when needed.

That women usually have no help at all, not even a cup of tea after a race nor, all too often, the basic and much-needed pre-race requirement of a loo, is farcical. We do not expect a changing room to be built at great cost for so few to use, but some form of sanitation should be made a priority.

"Men would be up in arms if treated like that and racing would very soon grind to a halt," declared trainer Gardie Grissell. "Some courses seem to regard women as a temporary phenomenon but this is simply not so. There may never be many, but those that there are should be properly looked after, not stuck miles away with no facilities."

Fontwell and Folkestone both use comfortable staff cottages but it is a long walk when carrying saddles to the weighing room or when trainers want to find their jockeys. Fontwell has drawn up plans for a ladies' changing room in its general re-building of the weighing block. Bangor, Warwick and Uttoxeter all have changing rooms and Plumpton was able to seal off a room-cum-passage, complete with mirror and wash handbasin, in the main block.

Happy memories were evoked when I rode at Plumpton. It was not my first visit to the weighing-room as I had once slept there for a week while at Pony Club Camp.

We enjoyed midnight feasts in the grandstand – and the boys also played truant from their quarters! This Sussex track, around which I remember racing on my own two feet, nestles beneath the South Downs. Thereafter it was always a favourite with me.

At Market Rasen women have the use of a room in the lads' hostel and a careful building programme over the years means that all the facilities are up to date and conveniently situated near the heart of affairs. In the North, Grade One tracks like Ayr and Newcastle, geared up to ladies' Flat races since 1972, have proper facilities. Kelso, where the corner of a room is cordoned off, is one of the few places which provides a woman to look after the lady riders. It is also a friendly and helpful course.

Taunton provided almost the opposite to a helper for the lady jockeys when, allocated the large first aid room (once again! Not to be recommended psychologically) the two riders were told to change in a small area behind a screen. A female first-aider was posted to see that they did not spread out and when Diana Grissell's sister came in to help her, she was promptly refused admittance. Angry, for this is a time when a sister can be at her most sisterly, Gardie Grissell tried to make the woman see reason. When she threatened to report him to the clerk of the course, Mr. Grissell quickly summoned that official and Cmdr. E. W. Sykes soon had things sorted out equably.

In the medical room at Towcester one day where no screen was available, jockey John Burke was curled up in agony with a dislocated knee. He groaned disarmingly as Di Grissell prepared for her race.

"The doctor was examining him and I just had to turn my back and get on with changing," she recalled.

Riding at Newton Abbot, Sue Horton dashed back to the ambulance room/ladies changing room for a hairnet and encountered a stark naked, prostrate, injured jockey. It was Johnny Francome.

The pioneer women jockeys took such things in their stride and will look back on them nostalgically when their needs are eventually met in more sophisticated fashion.

The men came in for their share of leg-pulling and a cheerful ribaldry soon grew up between male and female jockeys. If racing TV could home in on some of the good-humoured banter at the start, or even when a race is on, it would be an enlightening experience . . .

CHAPTER SEVEN

FACING THE RISK

Two years after women appeared on the NH scene, their riding generally accepted and no longer considered newsworthy, a great many racing men – and women – still did not approve. This was not so much a question of hurt pride or deflated egos but a genuine feeling of anxiety about women's welfare. No right-minded male jockey wanted to have his horse land on a fallen female or see her kicked along the ground like a football. A few feared for the safety of other jockeys but this worry soon proved unfounded. With the introduction of A and B permits, novice riders of either sex could no longer ride against professionals anyway.

Of course NH racing is a dangerous sport, just as it is tough and challenging; but we can, after all, get knocked over crossing the road or hurt in a car crash, without any of the excitement.

The lure of racing gets in the blood; there is nothing to compare with the thrill of steeplechasing, of meeting a fence plumb right and soaring over it on a horse full of running, of sitting on a tight rein approaching the last while those around are scrubbing. These are the tingling sensations the grandstand jockey can never experience!

The risk of permanent injury or disfigurement is the greatest fear and, without dwelling on the subject, it does need to be weighed up, though too much thought of injury can be a sign of failing nerve. "A fall is an 'orrible thing" as Jorrocks said and they are an inevitable part of NH racing, even from "patent safeties". But it is the amateur's prerogative not to ride the rough or dangerous horses. They will get falls enough without looking for them.

No amount of money or other assistance can compensate for permanent disability resulting from injuries, but can play an important part in easing the suffering in a practical way. When Tim Brookshaw and Paddy Farrell broke their backs in the 1963–64 season (Paddy from a fall in the Grand National with Border Flight) public response to an appeal for them was so generous that money was still coming in after the fund closed at £47,809. The surplus was used to found a new fund called the Injured NH Jockeys' Fund, standing initially at £6,000.

The original Trustees were Clifford Nicholson (chairman), Edward Courage, Fred Winter, Wing Cmdr. P. D. O. Vaux and the Hon. John Lawrence (now Lord Oaksey). Stan Mellor liaised between the fund and the professional NH Jockeys' Association while Col. J. A. T. Barstow, a solicitor and cousin of Lord Oaksey, was the first secretary. They met four times a year and in those early days carried out most of the case-visiting themselves. Mr. Courage, that great supporter of NH racing as breeder, owner and permit trainer of the lion-hearted Tiberetta/Spanish Steps line, was paralysed by polio just before the last war. He was therefore of particular help to paraplegic cases.

In 1969 Mrs. Susan Mills, a relation of Col. Barstow, was "press-ganged" into becoming an almoner, a voluntary role she still carries out conscientiously with other help. She is capable, hard-working and compassionate with beneficiaries; determined when dealing with Government departments. In 1971 Col. Barstow was elected to the Board of Trustees and Mr. J. Richardson, his assistant for four years, became secretary, also acting as an almoner. There are no full time employees and no fund-raising department as such. Mr. Richardson handles the co-ordination and administration of the fund, which is a registered charity, and Mrs. Barbara Bateson and her helpers organise and carry out the distribution of the Christmas cards, one of the principal sources of income since the start of the fund.

The standard of card is so high that it is hard to imagine Christmas without them, the appearance of the latest design being eagerly anticipated annually. Peter Biegel and Leesa Sandys Lumsdaine are two of the artists who have helped the fund with their designs. Fund raising functions and public donations make up the bulk of the remainder of annual income, and many give readily, knowing that their enjoyment of the jumping game relies on the risks undertaken by the jockeys.

In 1971 the fund was renamed the Injured Jockeys Fund, so incorporating Flat race jockeys, a move which resulted in greater general support from the Flat racing fraternity. Amateur riders of either sex are covered by its terms (no women jockeys as yet have become beneficiaries) but falls sustained point-to-pointing, as a general rule, are not. The Trustees can make exceptions.

One of the fund's first tasks was to track down and assist older jockeys injured before the National Hunt Accident Fund, started in 1924 for racing risks only, began helping those permanently disabled from the early 1950s. The Horserace Betting Levy Board ran an official compensation scheme (now administered by the Racehorse Owners Compensation Fund) but those cases which, for one reason or another, are ineligible now come under the wing of the Injured Jockeys' Fund – such as a jockey badly hurt in a car crash, and general assistance beyond mere compensation.

The scope of its work is wide and varied – not only paying weekly grants but in arranging low mortgages and giving loans to enable jockeys, who have been forced to quit, to set up small businesses or to train for a new trade or profession. Where the need is justified, a car, TV, holiday or, in certain cases, help with school fees are given. A paraplegic case had a purpose-designed bungalow built for him. Medical treatment, the advice of specialist consultants and the expenses involved with attending the Medical Rehabilitation Centre are also paid.

Since its inception in 1964, the IJF has distributed well over a quarter of a million pounds to more than a hundred jockeys and their dependants. But though there is a credit balance of some £100,000 in the fund, this produces investment income of only about £6,000 a year.

With weekly grants of about £450, some £25,000 ready cash is needed annually, a big discrepancy which leaves the Trustees in the unenviable position of knowing that continuance of the good work *depends* on the continuing goodwill of the public.

Mrs. Lester Piggott and Mr. R. J. McCreery have joined Messrs Courage and Winter, Lord Oaksey, Col. Barstow and Wing Cmdr. Vaux as Trustees, and their link now with the Jockeys' Association is secretary Major Peter Smith.

With such a team, jockeys' interests are certain to be looked after in the best possible way. As the growing public enjoys and appreciates the sport and the men and women who make it possible, they will surely continue to support such a worth-while cause. As Lord Oaksey says in his annual report, inflation, hard times and rising costs do not, unfortunately, lessen the risks run by men and women who ride horses with skill and courage for public entertainment . . . and as long as they continue riding horses at racing speed falls and injury will happen.

The few girls who take up NH racing professionally will, with sufficient rides of varying qualities, inevitably face an increased risk of injury. Only Sarah Eagleton, based with Arthur Stephenson in Co. Durham, and Mandy Rowlands, with trainer Peter Green near Nottingham, held professional licences in 1976–77 and at the start of the 1977–78 season they were joined by Vera Sharpe, who worked for Sussex permit holder Denis Browning. These girls just rode one or two of the horses they looked after professionally, so that there was little difference between them and the amateur lady riders – except that they would receive a riding fee.

Sarah Eagleton was mad on gymkhanas as a youngster, buying and selling foals and building up a friend's riding school. She left a point-to-point yard where she was head girl when the Sex Discrimination Act came in because she would no longer be able to point-to-point. She joined Arthur Stephenson's big stable. Sarah believes that if she were younger, leaving school and signing on with a trainer, her opportunities would be greater. Until more young girls are taken on in the same way as lads, she sees women's role as a minor one.

Says she: "I feel equal to a lot of lads and I will just keep trying to get rides. It may be easier for amateurs to get rides because there is no fee involved."

Arthur Stephenson gave her a few novice hurdle rides on young horses she looked after, her first being on Granton at Carlisle on Easter Saturday, 1977.

Mandy Rowlands also had two novice hurdle rides that year for her guv'nor. She rode Bird's Well at Leicester in January, finishing last in ground so heavy that ten of the 23 runners pulled up. Then she rode Catandale at Nottingham in March, pulling up before the second last after making the early running.

Vera Sharpe began by joining Charing, Kent trainer Robin Blakeney, spent a useful spell at Frank Cundell's where she learnt much schooling alongside Bob Davies, and spent a year with Miss Auriol Sinclair before joining Denis Browning. Denis helped her to obtain a professional licence and gave her her first ride in public on Rossula in a selling hurdle at Fontwell in October 1977. As Mrs. Alford, Vera had successfully point-to-pointed, when this was permitted for stable girls, winning on Spifkins on her debut. This was to my cost. She caught my mount, Comci Comca, in the final strides.

It might be imagined that few women would continue racing after they have had children but, in practice, this seems seldom to be so. I have no time for the women's libbers who try to make out that men and women are *equal*. Of course they are not, and *vive le difference!* The effect of child bearing is one thing male jockeys do not have to consider! As for equality in *ability* this is not a question of gender but of the individual.

The question of physical strength crops up time and again and it is true that, pound for pound, the average man is stronger than the average woman. But a woman with the courage and determination to succeed in racing has what I call moral strength. Then there is the quality of knack. Some horses unquestionably run better for a girl, a point which will, surely, in time, persuade trainers to try girls on horses sickened by punishment on the racecourse.

There are two types of strength needed in a race assuming that professionals and amateurs of either sex are already physically fit. First, there is that needed on the horse which pulls hard in the early stages. Plenty of men have been "carted" in their time, but the knack of holding a puller is just as important here as physical strength. A good horseman or horsewoman can succeed where a strong but perhaps ham-fisted pilot may not. Hence a woman will not necessarily be outgunned in this situation. Then there is the strength required in the closing stages of a race, when a driving finish is battled out and a tired horse needs "holding together". Physical strength, fitness and racing experience play their parts here. It is the harder type of strength to muster and this could be where the average man has the advantage over the average woman.

One school of thought holds that racing is too masculine a sport for women, but when it comes to femininity, we have only to look at the girls who go steeplechasing and hurdling to realise – forgiving a certain amount of bias here – that they are not weather-beaten, horse-faced, tough old hags! Their features may be disguised beneath skull caps, goggles, back-protectors and grimacing faces in the thick of the race, but they come bouncing back with pretty smiles!

A few years spent visiting smashed-up jockeys in the Jockey Club-appointed rehabilitation centre in Kilburn convinced Dorothy Laird that NH racing was no place for women.

She said: "I have great sympathy with men who do not like racing against women, although women have done so well and I admire them tremendously."

The male jockeys hold varying views. Bob Champion makes a woman feel perfectly at ease walking out to the paddock, yet he wishes he did not have to ride alongside her. He believes the game is too tough for a woman, or *should* be (he does not like the idea of unfeminine amazons) and that their bodies are not the right shape for falls (I know to my cost that a chest-first fall can hurt a woman more!) He also feels that few women are competitive enough.

Josh Gifford, former top NH jockey then a leading trainer, was more forthright. His opinion of women racing?: "Ridiculous!" He could not reconcile himself to their presence at all other than, perhaps, on a selling chaser or hunter chaser, the family horse that the woman knows well. Most lady riders fall into this category and those who ride most are mainly trainers' wives or daughters with greater scope. The sight of women in novice hurdles is abhorrent to Josh Gifford, nor does he believe they have enough physical strength for the sport.

This latter view is shared by David Nicholson (his wife, Dinah, was successful on the flat) who holds other strong opinions too! He takes the opposite line to Josh Gifford, condoning women in hurdle races to a certain degree, but not in chases. It was Nicholson who trained Joan Barrow for BBC TV's documentary, The Big Time, a task he performed conscientiously and professionally without regard to his personal prejudices. When the filming commitment was over he gave Mrs. Barrow more hurdle rides – one of them another winner on Jackstones, hero of the TV film.

When Joan suffered a crashing fall on her own horse in a hunters chase at Ayr later on "the Duke" was able to tell her what happened and why, without having been there! "You must have been on the inside, going too fast and riding too long". He was dead right as Joan ruefully admitted. She had taken the inside, where the fences were bigger, because her horse, Philuminist, was "as safe as houses". She had tried to go out in front to avoid trouble, but the poor horse never saw the first in the rush and she had been crushed beneath him, dislocating her pelvis. She had not been thrown clear as a jockey riding really short almost certainly would have been.

The craze in general equestrian activities for riding long with thighs embedded in deep, knee-rolled saddles, has probably caused an increase in serious falls, and for this reason: the rider who qualifies his point-to-pointer out hunting with his knees up may, at times, be looser in the saddle; but when it comes to the serious fall he is more likely to be shot clear and suffer less than the rider who is so much part of the saddle that he cannot get away in time.

It is the eventing style, or "fanny crouch: legs back, bottoms up, all bust and backside" as David Nicholson so delicately puts it, that he particularly criticizes amongst women in NH racing. He also feels that a woman does not have the physique to hold her own against the top professionals while conceding that they can be capable of becoming "reasonably competent" in the hurdling field.

"But what I said all along would happen, that a woman would not get away from a horse on the ground, is what happened to Joan Barrow."

Talking of injury, he said: "There will be an uproar when a lady-ridden horse comes into the last at the NH festival at Cheltenham and both are laid out cold. People will demand to know who allowed women into NH racing in the first place. Hurdles need not be too bad but chasing – never! I hate the idea of a woman being smashed up. It needn't be in a fall either. In the heat of the moment with the taps turned on, a professional isn't going to politely ask a girl to move over; he'll chop her into the rails smashing her leg and only stop to think of her sex after it's all over."

Yet Stan Mellor, whose career ran along somewhat parallel lines to David Nicholson, and who rode the NH record of over 1,000 winners, holds a contrary view. Now a successful trainer with his wife, Elaine, the well-deserved 1977 ladies Flat race champion, he believes that competent women should have been allowed to ride over fences long ago, putting the criteria on ability. Nor does he believe it is so much a question of physical strength as of knack, views I heartily endorse.

Fred Winter raises no objections to women in NH racing and is mildly surprised that there are not more of them. Says he: "They have taken a sensible approach and have done very well. They've certainly been no trouble."

David Mould, who was not renowned for his love of amateurs during his race-riding days, simply feels that chasing is too dangerous.

"It's just as well I never wanted to race," said his show-jumping star wife, Marion, "it would probably have ended in divorce!"

Top jockey Jeff King is not over-enamoured with them, saying simply: "Women'll never make jockeys, but so long as they keep out of my way, let them enjoy themselves."

Though Arthur Stephenson has supported Sarah Eagleton and given her occasional rides in public, he feels that professional girls will remain few under NH rules. He reckons that they, simply, are not good enough.

Plate 3 An historic photograph, reproduced from the photo-finish film taken by Racecourse Technical Services. Diana Thorne made history in becoming the first woman to ride a NH winner in Britain – beating her own father by a neck – in the Nimrod Hunter Steeplechase at Stratford on February 7, 1976. Women may lack physical strength in comparison to men, but they can often win by sheer determination.

Plate 4 Nora Wilmot (*right*), a pioneer for women's right's in racing, talking to Anne, Duchess of Westminster, owner of the mighty Arkle.

Says he: "The proof of the pudding is in the eating and very few are racing professionally. Perhaps I will be proved wrong and there may be the odd, exceptional, case."

Lord Oaksey, former top amateur rider and leading racing journalist who brings such life to his reports, was not too happy, at first, about women entering the NH field: "I felt reluctant simply because I didn't like the idea of racing against them knowing they might get hurt – not because I was afraid they might beat me!"

He added: "Touch wood, this has turned out to be mistaken. My view now is that women who have gone jumping have been much nearer excellence than those on the Flat, because they already had experience of cross-country riding and point-to-pointing. They are extremely good horsewomen, and nobody has rushed in and been stupid about it – girls are less likely to go jumping simply for glamour seeking."

Top jockey Jonjo O'Neill accepts women jockeys in his pleasant Irish manner but with certain reservations. If he knows a rider is capable he is not too concerned but considers "weekend" riders dangerous. He thinks the new A and B permits are a good thing.

"I have come across a few bad 'uns but you can get them in men as well," he admitted. "I don't mind women in racing if they are crazy enough to want to do it, and I know a lot of them who are, in fact, stronger than some of the young lads riding."

It must be remembered that top flight professionals are racing virtually every day, often several times, for their living, while amateurs of either sex, by the very nature of things, are bound to be more occasional riders.

Women jockeys also hold divided views on their role in racing. Gillian Fortescue-Thomas has more experience than most of competing against men on equal terms through a successful motor racing career. She is another one who denounces "women's libbing".

"If you approach mixed racing with a sensible attitude, competence and confidence, the men will respect you."

Making the valid point that it is much easier to win a race on a good horse than to complete the course on a bad one, she believes that a woman's physical strength will run out before a man's.

"I can't see girls riding six races in an afternoon like some professionals; one horse might be a puller, another will fall, another will want 'scrubbing' round; but I think a girl can ride a good horse that she knows as well as any professional – and some horses appreciate girls more, they are often better or more sympathetic horsewomen."

She feels that women who shout about Women's Lib (and I know of precious few in racing) seldom go out and do things themselves. Women have the best of both worlds today, she believes; often quite true – and how nice it is! The stewards at Cheltenham and Ascot, for instance, are charming and go out of their way to look after the girls. Good old fashioned respect and courtesy make racing even more enjoyable.

Sarah Eagleton, one of the tiny band of professional girl NH riders, finds that she is "treated as a jockey on the course, and as feminine off it."

Nicky Lay, who rode in more NH races than any other woman in the first two years, with 43 of her 50 rides being against professionals, finds the attitudes of other jockeys basically acceptable, but that if any were "going to be a bit funny" it would be the younger ones. Established jockeys, fearing no threat to their living, are chatty and friendly with plenty of jokes to crack; but like many other lady riders, Nicky Lay finds amateurs "hell bent" on shouting at women during a race.

Diana Grissell finds it is the amateurs who keep swearing, but the professionals are helpful. She admits being anxious not to do anything wrong under rules under the beady, critical, eyes of a Press and public.

Sue Horton knows the meaning of serious injury. She broke her pelvis and vertebrae point-to-pointing but found the fences better built under NH rules – this is the case in most parts of the country – and she loves the atmosphere: "The jockeys are great fun. Jeff King never stops talking and joking all the way round. They don't mind women so long as they are sensible."

Rosemary White sampled a taste of sour grapes in a hunter

chase. A man who apparently feared his "rights" might be usurped not only swore at her "impertinence" for going for an open gap on the inside, but sent messages afterwards warning her not to try it again, without having the courage to speak to her himself!

There have been very few cases of "doing the dirty" on a woman, but it did happen to Nicky Lay, not from a jockey but an owner. Asked by someone she knew to ride in a three mile chase at Chepstow, Nicky readily agreed without looking up the form first, naturally trusting a friend. On the morning of the race she discovered the form read "refused, fell, fell." Not encouraging! Then, in the paddock, the head lad had a quiet word with her: "Try and keep him covered up, ma'am, this one might try and duck out; and don't pick up the reins in the paddock, just sit quietly."

As she was legged up, the horse bucked, plunged, then reared over backwards, giving Mrs. Lay no chance and hurling her to the ground.

Victor Lay was summoned from the weighing room: "Your missus is on the floor."

"She can't be, the race hasn't started yet!"

Shaken, the intrepid Mrs. Lay re-mounted, but down at the start the rogue reared over backwards again and was withdrawn.

To give him his due the owner, belatedly trying to do the right thing, assured the stewards that he would never run the horse again, and sent him to the sales, without reserve.

Victor summed up the episode: "It was really doing a woman down. I had that sort of thing happen to me as a professional jockey but I was being paid for it. An amateur of either sex rides for fun, they want to enjoy it."

Male reaction improved generally with time.

When I rode in a field of 23 at Stratford I assured other jockeys that my horse jumped well and would be kept straight. But milling round at the start one jockey looked at me seriously and said: "I should keep on the wide outside, dear." Others tended to ignore my presence but in later races I found the jockeys cracking jokes and quite at ease, with a mutual feeling of being at home.

They soon realised that few women would be riding rogues or creating nuisances. As Geraldine Rees said, she "wouldn't get up on any old lunatic for the sake of it. I don't want to go looking for a broken nose and no teeth. A woman has got her looks to think about."

This brings us back to the two basic objections to women riding over fences: injuries and strength. Lewes trainer Miss Auriol Sinclair said, "I wouldn't put up a woman however good she was because I don't think they are built for the falls."

Yet how marvellous it is that women are no longer denied the right to sample NH racing's thrills!

As for the future, well over 100 women have ridden under NH rules already and this number is likely to increase as up-and-coming riders join in, probably from the point-to-point field. The tiny professional group should swell as trainers become more prepared to give them a chance. In time one or two could make a name. Women's role may remain small but significant and their record in their first full season, 1976–77, speaks for itself. With 418 horses ridden by 85 women, 42 were winners, or virtually 10%. They notched up 28 seconds, 32 thirds, 28 fourths and of the remainder, 42 pulled up, 30 fell, two ran out, three refused, two were brought down, three slipped up, 11 unseated rider and the other 195 "also ran".

CHAPTER EIGHT

GILLIAN FORTESCUE THOMAS

WOMEN were inevitably in the limelight in their first (half) season, but their record was impressive. Gillian Fortescue Thomas became the first ladies' champion, scoring three wins and three seconds in between January and June 1976. It was a title which cut little ice at the time, and Mrs. Fortescue Thomas had to wait until December 1977 before being presented with her promised trophy!

Local donkey races first publicly revealed the inborn competitive spirit of Gillian Fortescue Thomas. At the age of 11, she rode with dash and aplomb at Headcorn near her home in Kent, beating the boys regularly. She trained and rode a white donkey named Omo to win the local donkey Grand National, a tough test over three-quarters of a mile and straw bale fences.

When there was not a four-legged animal to race, Gillian used her own two feet, running and hurdling with success at school standard. The donkeys were fun but she soon craved more and found her opening in point-to-pointing, starting on a schoolmaster, Greek Gambler, then taking a succession of extremely moderate "spare" rides for other owners. One she recalls wryly. After it had reared over backwards twice with her in the paddock, she was legged back into the saddle by her husband, Anthony, who, far from showing concern, was grinning from ear to ear.

This tall, bronzed farmer and successful show-jumper has a charming, phlegmatic nature which enabled him to take his wife's exploits in his stride. This was just as well when Gillian, finding insufficient edge in point-to-pointing, turned her hand to motor racing.

Again, it was not so much a question of loving cars or driving, but of burning competitive spirit. Gillian's good looks, her long, fair hair with a touch of auburn, attractive freckles, dark, wide eyes, brought glamour to the rough and tough scramble of auto-cross. She quickly climbed to the heights on grass, progressed to rally cross and then on to circuit racing, making an impact wherever she went and taking equal competition against men as a matter of course. Her first win in a circuit race was in a car owned by Anthony, who coupled as mechanic. Gillian's big break came when she performed a successful test drive for Ford and went on to win a rally championship for them.

A test drive at Silverstone against international drivers secured her a contract to drive professionally for Ford. It sent her to many parts of Europe and into all sorts of hair-raising incidents. About her numerous knocks she is reticent, having long learnt to live with them. Her career highlights include a world record attempt with a diesel car at Monza, Italy, and driving in a 24-hour saloon car race at Spa in Belgium, but perhaps her greatest moment was on her home track at Brands Hatch. Determined to win in front of the local crowd and, more important, to impress the professional talent spotters, her practice time was 2/10ths of a second behind an up-and-coming South African by the name of Jodi Schechter. It was Gillian's first season and it was a race in which all competitors had identical cars. Gillian led for the first nine-and-a-half of the ten laps, but had to give way to Jodi Schechter. She was disappointed by her "failure" but now, with the benefit of hindsight, she reflects that it was not too poor a performance!

When she and Anthony bought 110 acres and a decaying farmhouse in Dorset, the distance proved too great a barrier for Gillian to continue her professional motor racing career. Instead, the couple spent every spare minute doing up the attractive thatched stone house ready for housing paying guests in the summer. And they began taking in horses at livery for schooling, breaking, hunting – and point-to-pointing.

So Gillian found herself back with race-horses, schooling a young novice of Tim Frost's called Prince Rock in his early point-to-points, concentrating on his jumping education at home over poles. He was weak and under-developed as a six-year-old, but came out the following year furnished up and stronger to win all seven of his point-to-points. When Gillian dismounted after beating Horoscope in the three-and-a-half-mile ladies' event at the Heythrop, Gillian declared: "This is a Grand National horse."

Prince Rock was sold into Peter Bailey's yard and was soon successful under NH rules for his new owner, Michael Buckley. He made his first assault on the Grand National as a nine-year-old in 1977, falling at the 12th when close up.

After Prince Rock, Gillian became associated with Stanhope Street. By a stroke of good fortune, it coincided with the year women could apply to ride under NH rules.

Stanhope Street proved a bit of a puzzle. A promising point-to-pointer at five and six years, he won a few moderate races but was more often let down by his jumping. With maturity at seven, this problem was ironed out and he won seven men's point-to-points, trained by Anne Harden, and ridden by Barry Venn. At eight, he carried all before him, winning on all three of his appearances under NH rules. These were not novice events but included the Players Gold Leaf final at Chepstow and the Horse and Hound Cup Final Champion hunters chase at Stratford.

His Somerset owner, Mr. Bill Counsell, decided to have the horse professionally trained the following year and aimed at top races, so Stanhope Street left Anne Harden, a genuine, sporting housewife who had had well deserved success a few years earlier with Darnick Tower. The move to Tim Forster's proved a complete flop. Tim Forster is an expert at his craft with a string of big successes to show for it, but it seems that Stanhope Street, always a "bit of a character", detested the regimentation of a professional yard and showed it on the racecourse, only once making the frame – a moderate third.

Though top trainers have great skill, excellent facilities and the best of fodder, care and professional expertise available, one advantage remains with the "home trainer", the permit-holder or, in the case of hunter-chasers only, the owner or livery yard proprietor. This is the very smallness and informality of it all. Beautiful gallops and strict, efficient routine may be lacking, but compensation comes by the individual attention, devoted personal care and, above all, variety. A horse who has become stale in training will often be revitalised by such a yard, not knowing which way he will head tomorrow . . . the woods, the forestry, the sea. But hardly ever for gallops. Hunting is often the most beneficial remedy of all and despite risks of injury, the new zest and *joie-de-vivre* it gives a bored horse more than pays off.

Such was the case with Stanhope Street. He had shown his appreciation of a woman's care when with Anne Harden. So, as the horsebox carried him up the long concrete drive to the quiet, pretty setting of the Fortescue Thomas's South Buckham Farm for the start of the 1975–76 season, the Dorset countryside rolling away into Somerset, he almost audibly sighed with pleasure.

His come-back was as complete as had been his failure the previous year. Back to point-to-points after a hard season's hunting with the Seavington, he twice beat the useful Marshalsland in ladies' races before being earmarked for the Cheltenham Festival. What an occasion for a woman's first ride under NH rules!

Gillian was not the first woman to face Cheltenham's stiff fences (Jenny Stamp had that privilege on the opening day) but it was quite an ordeal even for a girl with nerves of steel. And so it proved in the race. For many punters in the stands, the sight of Stanhope Street trailing the field of 13 for the Cathcart Hunter's Chase, the final "get-out" of the three day meeting, smacked of lack of nerve. It was, of course, Stanhope Street's first reappearance on a pukka course since his baffling loss of form the previous year, and it now appeared that his state of mind was not yet tuned to tackle the big scene.

He was lacklustre all through the first circuit and tailed off by the seventh fence. But Gillian resisted the temptation to hit him: "I had to sit and suffer, feeling a complete idiot," she recalled.

She kept niggling at the horse on the second circuit but was still a distant ninth turning into the straight when, suddenly, the horse took it into his head to go, making phenomenal headway. Jumping the last well behind, he passed half-a-dozen runners from there to the post, getting the better of Tartan Slave by a neck for second place but finding Mickley Seabright too impossibly far ahead to catch.

There was no question mark about his next performance. With his measure better sized-up, Gillian established more of a partnership in the Merlin Hunters Chase at Ascot. She had, meanwhile, scored most confidently on him at the Heythrop Ladies race accounting for the previously unbeaten point-to-pointer, Zanetta. At Ascot Stanhope beat Playbill and Nicky Brown half a length and 25, drawing clear with the John Sharp-ridden runner-up from three out.

Now Gillian was really at home on Stanhope Street. Her motor racing experience almost certainly helped, enabling her to separate her mind from her body picking out rivals' colours, noting their progress and judging pace but still maintaining total unison with her horse. This sort of concentration distinguishes the good from the ordinary.

Next time out she went under by a length and a half to False Note in a three miles five furlongs event at Warwick and this time Gillian did blame herself for lying too far out of her ground – especially as she subsequently beat False Note convincingly in the Horse and Hound Cup.

The champion hunter chase was a fine climax to the season. She went under with all honour, to Otter Way, a class horse of Gold Cup calibre, trained in similar "hunting" fashion in Devon by his breeder and permit-trainer-owner, Mr. Oliver Carter. Otter Way's warm-up for Stratford had been to win the Whitbread Gold Cup!

Sadly, in 1976–77, Stanhope Street flopped dismally, failing to gain a single place in five outings. Gillian believed it was because he injured a foot in the winter causing him to miss much of his hunting tonic. He was then taken straight to a racecourse instead of being kidded along with some easy point-to-points.

But with three wins and three seconds in 1975–76 in hunter chases on Stanhope Street (and an unplaced ride on Field House in a hurdle) Gillian Fortescue Thomas became the first NH female champion, for which *The Field* had promised a trophy.

Since Muriel Naughton's ride on Ballycasey at Ayr on January 30, 145 horses had been ridden by women, many of them in hunter chases. Mrs. Fortescue Thomas was the only woman to gain more than one success in that time. Seven other women each rode one winner. Seven of the ten lady-ridden winners were for hunter chasers. The record was impressive: ten wins, 15 seconds, 13 thirds and 12 fourths. Fourteen fell, but only three were recorded as "unseated rider", proving that women were not falling off all over the place – though some jockeys who are decanted find it recorded as a fall. Race-readers are amazingly accurate with the "also rans", a very tricky job, and they do not necessarily rely on a jockey who might want to save face by saying his mount fell, though he was, in fact, unseated. But mistakes do sometimes occur.

CHAPTER NINE

NICKY LAY

THE woman who made the biggest initial impact was Nicky Lay, both in terms of number of rides and proficiency.

Victor and Nicolette Lay moved to the stables at Broughton, near Banbury, soon after their marriage in 1973 and after Victor's father, Ben, gave up training to retire in Ireland. Victor had been a good NH jockey but his career was shortened after he had been badly smashed up at Fakenham. Though he took over his father's training licence, things were not handed to him on a plate. Only one horse came with it, old Headlight, winner in his prime of ten races, all with Victor up. By Coalition out of a mare called Beacon Light, he would give Nicky a good ride in a chase when he was 15. A painting of him takes pride of place in their home, which is set conveniently beside the stables.

Nicky grew up with horses and could ride almost before she could walk. Her uncle, Henry Boswell, was a trainer and her mother kept the odd point-to-pointer. By 13, Nicky was all set for her first ride in a point-to-point on an old schoolmaster when, in 1962, the Jockey Club introduced its minimum age of 18 for lady riders in these events (it was altered to 16 years for either sex after the SDA). So the tall, slim blonde had to wait until she was 18, filling in her time by show-jumping and a much-detested year in hairdressing.

Her expected mount went lame and so Nicky was 19 when she at last made it to a point-to-point course – with disastrous results. Her four rides produced four falls, including one at the first fence at the Heythrop from whence she was conveyed unconscious to Oxford Hospital. Undeterred Nicky began the next season more determined than ever to get round. It was not plain sailing. She broke her leg in a hunting accident in November and did not have the plaster removed until the following April. Yet only ten days later she rode in a race – and finished third.

By the time the SDA came into force Nicky was a polished and proficient point-to-point rider. When she first mentioned the possibility of women riding under NH rules, Victor's reply was "rubbish". But when it materialised, he was the first to encourage her to apply for a licence. A short time before, Nicky and a few other point-to-point girls had planned to organise a petition to the Jockey Club calling for a ladies' hunters chase.

With the new law on the statute book, one of Victor's owners, Mr. Toddy Jefferies, said he would like Nicky to ride his horse, Abbey Farm. On February 18 Nicky found herself cantering down to the two mile hurdle start at Towcester, against 11 professional jockeys. It was Division 4, Part 1, of a four-year-old novices hurdle and the starting price of 50-1 reflected Abbey Farm's chances.

Nicky had walked the course with Victor in a downpour and he advised her how to ride the race. Apprehension set in, not so much about the race itself but because it was against professionals and Nicky was afraid she would be unpopular with them. But Ian Watkinson, the only one she knew, was reassuring. It was the eighth of the nine races on the mammoth card swelled by the glut of novice hurdlers, an agonisingly long time to wait.

The runners set off at a cracking pace and down the hill they were really "dinging on". On a moderate horse Nicky could have been forgiven for dropping back, but to her credit she kept her mount handy and three from home the chestnut struck the front. It was not an effort he was capable of sustaining, but although he weakened once the pressure was on he was not disgraced to finish seventh. Neither was his jockey disgraced. The professionals had treated her as just another jockey in the race and she had kept a straight course and cool head. She soon preferred riding against professionals though some of the younger ones appeared slightly resentful. But David Sunderland, who lost a good many of Victor Lay's rides as a result of Nicky, bore no ill-will. This would have conflicted with his cheerful Irish nature. Worst were the male amateurs, often shouting and swearing unnecessarily during a race, a view supported by many other women riders.

By the tail end of the season Nicky had gained valuable experience in hurdles and chases and had the bit firmly between her teeth by the time she had the ride on a tiny chestnut mare called Berostina in an amateur riders' handicap hurdle at Hereford. It was Spring Bank Holiday and there was a glut of meetings.

Berostina was trained by David Hanley who held a restricted permit, confining his horses to amateur races only. Luckily this severe restriction was dispensed with the following season.

An unrestricted permit was not allowed until the restricted holder had won a race. It is not easy to win races at the best of times, and most amateur events are highly competitive, patronised by top trainers. Many are races of high value. So it was an almost impossible position for the farmer or similar permit-holder who liked to train one or two horses himself for pleasure, not business.

The restriction was designed to reduce the size of novice hurdle fields and to ensure that licensed trainers did not have their livelihoods put at risk. In effect the novice hurdle fields became no smaller, and few permit-holders would have put horses into training had they not possessed a permit: to them it would have meant losing the rewarding personal touch. Instead they would have been lost to racing altogether.

At 15hh Berostina did not have size on her side but she was remarkably well-bred. By Ribero out of an Acropolis mare, she had been beaten as a two-year-old by only half a length to Dibidale, luckless heroine of Polygamy's Oaks when her saddle slipped back to her stifle yet she still finished third.

A campaign in America at three followed by hurdling in this country, a switch to the Flat, then, still with no rest, back to hurdling was the sort of programme which would have killed the spirit of many. But tiny Berostina had a big, courageous heart, that intangible but greatest of all assets in a racehorse. When Nicky first rode her at Worcester there were 30 runners and, after five previous outings unplaced, she started at 50-1, as did Indian Delight ridden by Judy Goodhew. Mrs. Jenny Owen and Scupper were at 33-1. Indian Delight pulled up and the other two were unplaced, but Berostina had given Nicky an encouraging run, only losing her place after she was badly balked.

So if hopes were not exactly high when Nicky set off for Hereford on the Bank Holiday she felt the mare would go well.

It was a two mile amateur handicap with 12 runners. Nicky Lay, Mrs. Owen and Mrs. K. Huntley Jones on Honey Lover were three of the five to carry the minimum of 9st 7lb and they were among the rank outsiders. But this was Berostina's day. Nicky Lay rode a grand race on the mare, keeping her handy and producing her with dash, driving into the last and riding her out up the run-in, wearing down Ten Knots ridden by Shaun Parkyn till, on the line, she had half a length to spare.

It was a great moment, worth celebrating in style but that had to wait. Horse, jockey and Victor Lay had to hit the road again for the long journey south-west to Devon and Exeter where the brave mare was declared to run again, on hard ground, the next day. She emerged remarkably fresh from her exertions but this time it was a novice hurdle, for which she had to carry a penalty after her success of the previous day. She started at 5-2, second favourite of six for the testing course, with top weight of 11st 10lb. Once more the mare ran her heart out, disputing the running to halfway and then taking up the lead. Half way up the run-in it still looked as though she might hold on but the strong challenge of the favourite, Somers Glance (C. Watkins) prevailed within yards of the line giving that horse a narrow victory.

It was a short summer for Berostina, but she was used to working hard for her living and early in the 1976–77 season she ran another fine race at Worcester on August 7th, starting at 20-1. Making every yard of the running, Berostina made a mistake at the last and was joined by Mark's Boy. The two were locked together up the run-in but it was Mark's Boy (P. Steel) who got his nose in front. Berostina was unplaced thereafter and was eventually sent to Ascot Sales without a reserve. To the Lays' bitter regret she was virtually "given away". They had been unable to go because of racing at Worcester – ironically that was cancelled when they got there.

Someone bought a sound, tough, game little winning mare for a song – and the Lays hoped she did not end up in a dog-meat can, the fate of so many cheap lots as the value of meat soared. It is wiser to put on a reserve just above the current maximum meat price or, in the case of a horse with little worthwhile future, to send it to a reputable slaughterer and prevent undue suffering. Otherwise, with our sadly inadequate or unenforceable laws, there is evidence that horses are sometimes shipped to the continent without food, water or suitable supervision.

The 1976–77 season was the first full one open to women riders and confirmed Nicky Lay's place in the forefront of women NH riders. She notched a remarkable 50 rides, a high number in one season for any amateur, and by the end of it, had ridden more winners than any other woman. What made her achievement more notable was that she won on four different horses. She was not gracing the winner's enclosure by virtue of one good horse.

The reason for Nicky's successes soon became clear. She rode stylishly, and used her head and was more adept than many amateurs at riding a finish.

But her real forte lay in riding into the last, where so many races are lost or won and where the men are truly sorted from the boys. Mrs. Lay throws her heart over as fearlessly as many a great professional.

One of her favourite rides was on the former top flight chaser, Chatham, then hunter-chasing for Mrs. Hilary Trigg. Though the best she did was to be second to that year's leading hunter-chaser, Remigio, Nicky Lay found Chatham in a class of his own compared with her usual horses. He gave her a Rolls Royce of a ride. Another hunter-chaser gave her her next win. Second on Caille to the Thornes' Air General, she won a hunter-chase on him at her favourite Devon and Exeter, a course where she always found the people friendly and welcoming. She beat the joint favourite Conchita (K. Bosley) and Devon Spirit (K. Pook) by riding the perfect waiting race, coming with a run from two out to lead on the flat. Mrs. Fizz Chown, a former ladies point-to-point champion, associated with such good horses as Royal Charity and

Headsprite, was unplaced on Silver Plate. Mrs. Chown, one of the most competent women riders, three times finished second on Cherry Blossom in hunter-chases in 1976.

Hereford was another of Mrs. Lay's lucky courses and there, on Bank Holiday, June 6, she notched her last success of the season, this time on a mare called Gentle Rose, for Michael Oliver. The horse had run 13 times, mostly unplaced, but once the ground firmed up and the better opposition bowed out, she gained a second and a third before winning in Mrs. Lay's capable hands, beating the 11-10 favourite, Swift Answer (Ron Hyett) by four lengths.

But Nicky had been seen at her best on Easter Monday riding another mare, the 11-year-old chestnut, Current Romance. In her younger days the mare had won the County Hurdle at Cheltenham, but had then been put unsuccessfully to stud. Nicky got her through Patsy Smiles (formerly Holland) vice chairman of the LJA, who had given up race-riding after her marriage to Peter Smiles, stewards' secretary and subsequently Director of Racecourse Security. Nicky Lay tried point-to-pointing the mare but she proved too hair-raising, jumping diabolically. So she switched to hurdling. When Current Romance appeared at Newton Abbot for the Easter Meeting she had not run for three years, as far as the general public was concerned. In reality she was fit from her efforts between the flags.

So Nicky Lay again found herself riding a rank outsider but though sticking to their no-betting rule, Victor did tell a few of his owners that she should go well. It was a good class three-and-a quarter-mile handicap hurdle. Nicky rode at her customary bottom weight – and rode brilliantly.

Victor was in the stands with one of his favourite owners, the elderly Mrs. W. H. Mowlem. He watched his wife intently as she rode to orders beautifully, well in touch half-way. Before the last she made rapid progress, produced her mount with a great run and flew the last, wearing down the leader, Sea Emperor (Clive Candy) on the run-in to get up by a head.

Victor was so overjoyed at her performance that he promptly embraced Mrs. Mowlem and kissed her. "I think I was more

Plate 5 Mrs. Nicky Ledger in action on Cedras at Fontwell. Nicky relished the speed of hurdling and won a novice race riding Bujoji.

Plate 6 The author, whose entry into racing was due to Parliament's enactment of the Sex Discrimination Act which allowed her, and other women like her, to sample the thrills and spills of NH racing.

Plate 7 As many married women participate under NH rules as do single, and husbands play an important part in encouraging their wives. The author's husband, Tony, on Log, the first horse the author rode under NH rules.

Plate 8 Vera Sharpe rides as a professional for Sussex based trainer Denis Browning.

excited about that than my own first winner," he exclaimed.

Once more there was little time for celebration. With runners next day at Uttoxeter, the box turned round and was soon trundling north-east.

Early in 1978, Nicky was brought down in a point-to-point at Cottenham. She broke an arm, some ribs and injured her knee.

CHAPTER TEN

JOAN BARROW AND "THE BIG TIME"

"THERE's something moving through the sheep on the far bank!"

John Barrow grabs the binoculars, which are never far from his side and raises them to his eyes: "It's a long dog!"

Strangely, he is not concerned about the sheep – they are not being harassed – but his beloved hares. This is coursing country and John guards the hares on his Cotswold hills as jealously as he cares for his lurchers.

His farm lies down a twisting, narrow, typically English country lane, nestled in a valley. A couple of horses graze in the paddock by the tree-lined drive to the creeper-clad, rambling old stone farmhouse. Two toddlers rush in and out with carefree abandon, playing with the dogs – there are a couple of hound puppies being "walked" and a spaniel as well as the lurchers. Down by the farm buildings some summer visitors are putting away the skewbald pony and trap, the reins having just been handed over by Joan Barrow, her short, dark hair bobbing as she breezes up to John holding the binoculars. It is only three weeks since she discarded her crutches, but already she has ridden out at David Nicholson's place.

"That's a winner every time we don't know how to entertain some-one," she says with a smile towards the skewbald.

That pony and trap, ambling peacefully along the lanes, became familiar to thousands of British TV viewers. The scene was featured at the end of the racing programme in the BBC series "The Big Time". This feature, compered throughout by Esther Rantzen, covered a variety of subjects with one basic theme: an ordinary, unknown person trying his hand at the top of his chosen sphere, be it cooking, conducting or racing.

The preparatory work and build up on each subject was meticulously recorded by Miss Rantzen and her team. Joan Barrow, mother of two and farmer's wife, would, in her own words, have been still "plodding round in the local point-to-points" but for "The Big Time".

It began when she finished second in her hunt members race at the nearby Heythrop point-to-point at Stow-on-the-Wold. Up came BBC racing commentator and ex-jockey Richard Pitman, looking for likely candidates for the series.

Half-a-dozen were chosen, mostly lesser known point-to-point riders from that area and who, at the time, had not ridden under rules. A gruelling interview in London followed, and there can be little doubt that the BBC found exactly what they wanted in the delightful personality and completely natural ways of Joan Barrow. What followed did not spoil Joan at all. She endured the intrusion into her private life with dignity and, before long, became accustomed to cameras whirring inches from her face – though she confessed she did not enjoy the interviews. But the public had to be shown what was involved and just what it means to be a woman who is determined to succeed in what, until recently, was a man's world.

Such personalities as Richard Pitman, Graham Thorner, Peter O'Sullevan – the greatest commentator of them all – and Sue Horton gave their views about women in racing and told gruesome tales of horrific falls and injuries. But the main theme was the preparation and training for the selected race and this covered a period of several months. For this, Stow-on-the-Wold trainer David Nicholson was chosen as mentor. A better choice could not have been made than this renowned perfectionist who left no stone unturned, no comment unspoken. The "Duke" never flannelled. He was known to have just three dislikes in life: amateur jockeys, women jockeys and amateur women jockeys. If Joan did something incorrectly he told her so, bluntly, and she took any amount of "stick" unflinchingly, especially in the early days, when even her clothes for riding out were criticized. Half measures were not good enough for a professional like the "Duke".

Her first morning's canter was full of mistakes but Joan Barrow was a willing and dedicated pupil. Her improvement, as weeks went by, was evident to all watching the TV programme.

Joan had always longed to race but money was limited and she made do with one moderate point-to-pointer. She and her husband, John, once rode together in a hunt race but he had something of a weight problem. This he blamed on Joan's excellent cooking which, like everything else, was not allowed to suffer interference from her TV commitment.

John Barrow, a former Master of the Eton College Beagles, preferred "doing" to watching and spent more time coursing, shooting and hunting when overweight prevented him racing. He liked watching Joan ride but was content to leave after her race. He did not object to the filming – for which, incidentally, only their bare expenses were covered – and he kept his sense of humour. There was little danger of Joan becoming too "horsey".

Joan loved every minute of the training, especially as she became more and more involved. She owed much to her "mother's help", Sheila Johnson, without whom she could not have left the children every day to ride out, run, cycle and take a full part in her training programme. Every few weeks she would be filmed at these activities. Luckily for Joan, Sheila Johnson appeared equally at home driving the tractor, feeding the dogs or whatever else she was called upon to do at short notice.

Once Joan had been chosen for the film part she was asked not to ride under NH rules until "The Big Time" crew were ready. They wanted the filmed race to be her first – and they wanted it to be against professionals. Here they encountered a problem as the licensing stewards considered Mrs. Barrow too inexperienced. They would allow her to ride in amateur races only, they said.

Esther Rantzen, not easily dissuaded, went pleading to Portman Square herself. To no avail! By this time the stewards were, possibly, a little wary – fearing a publicity stunt with some film star posing as a jockey! But they assured Miss Rantzen that a steward's secretary would watch Mrs. Barrow to see if she could be considered competent enough to ride against professionals later. But at the end of the day, when the film was shown, the stewards felt it had done nothing but good for the cause of jump racing.

They wrote to Miss Rantzen, congratulating her upon the high quality of the whole feature.

It was becoming a do-or-die effort for Joan. If she failed, it would be her last chance. A man like Nicholson was unlikely to give her a second go and anyway those cameras would be rolling. Joan's nerves were now beginning to tingle, but this gave her the extra adrenalin to cope with the situation.

The chosen horse was Mrs. R. Heathcote's Jackstones, a five-year-old bay gelding by Bargello. The final "no" having been given to a professional race, it was to be an amateur riders' event at Ayr, the Ayrshire Yeomanry cup novices hurdle, run over two-and-a-half miles on Monday, October 11, 1976. This was a happy venue for Joan. She was born and bred in Scotland and her parents farmed not far from the racecourse so they would be there on the big day. Jackstones also ran at Ayr in a two-and-a-half-mile Philip Cornes Novice hurdle qualifier on the Saturday, finishing second with R. Dickin in the saddle.

Once on the course Joan did not have a moment to herself – at just the time she would have preferred to be left alone. David Nicholson showed her all she had to do and walked the course with her on the morning of the race. The inevitable cameras were in close attendance as he instructed her, in minute detail, how to ride the race.

"And", said he, "don't look round under any circumstances or I'll ping you off with a 12 bore!"

She was seen being escorted to her changing room, shown weighing-out, putting on a hairnet and making final touches to her simple make-up, details no male jockey ever had to worry his head about! Esther Rantzen asked how she felt. Joan's reply was simple: "Sick!"

Cameras followed her into the paddock as she walked out with the other riders, all men, as she met her owner and had her last chat with David Nicholson. Now the moment had come. It was all up to her now – and to Jackstones.

There were 13 runners and after his good run of the Saturday Jackstones started favourite at 2-1, followed by Spartan Sandal at 5-2. The runners circled at the start behind the tapes for those interminable seconds while last minute adjustments to girths were made and the starter checked the roll-call. He called them into line and Joan niggled for position. A clean break was imperative.

As the tapes shot up she got off to a flyer, actually ducking the rising tapes and quickly settling her horse into his stride. The first flight flicked safely by and Jackstones moved into a narrow lead, accompanied by Bar Haze. After half way her lead increased, but Jackstones met a flight wrong, momentarily loosening Joan but not in the least checking his stride.

With two to go she was ten lengths clear. In the stands, the crowd began to hum and David Nicholson permitted a small smile to crease his face. Joan's parents, Ken and Tina McCall, were jumping up and down excitedly while John Barrow holloaed at the top of his voice oblivious to curious stares around him. Seconds later he was joined by enthusiastic cheering from the crowds.

The runners were heading for the last. Jackstones was 20 lengths clear and he had only to jump it to win – the others could *never* catch him now! He rose to the occasion and so did Joan, flying it as foot-perfectly as the first. On the run-in he increased his lead to an incredible 30 lengths.

The cameras kept going with the crew scarcely able to credit their luck.

"It's a dream come true," said David Nicholson.

"I can't believe it", panted the elated Joan.

Said the steward's secretary: "I can't say myself if she will be allowed to ride against professionals, but I will be forwarding my report. She doesn't seem to have done anything wrong." A masterly understatement.

Cheers from the crowd, hugs from parents and friends, a kiss from husband John, and Joan, still numbed by it all, unsaddled in the winner's enclosure and weighed in.

The official result: 1 Mrs. R. Heathcote's Jackstones (Mrs. J. Barrow), trainer D. Nicholson, by 30 lengths at 2-1; 2 Spartan Sandal (A. Eubank); 3 Tamieshanter (R. Page).

The unofficial verdict: a fairy tale.

Joan Barrow won again on Jackstones in a three mile handicap amateur hurdle at Chepstow on November 12, beating Semi-Colon by eight lengths and starting a 7-1 fourth favourite – and she got her permit to ride against the professionals. This she did – on Jackstones again – in the final of the Philip Cornes series at Newbury on March 5. This hotly competitive contest was won by The Dikler's half-brother Kas, Jackstones being somewhat outclassed.

In May Joan revisited Ayr, this time with Philuminist, for that ill-fated hunter-chase. She was taken to Ayr Hospital with a dislocated hip, followed by a spell nearer home at Oxford.

Everything went wrong from the start. The horse panicked in the box, which Joan had to drive all the way to Scotland the previous night. Next morning the box had a flat battery, preventing her starting for the course in good time. Philuminist, normally a calm character, was ashake before the race.

There were eight runners, three ridden by women and it was one of Scotland's leading lady point-to-point riders, Gillian Minto, who won on Cool Thrust, beating Miss Ailie Nisbet on Of Course by three-quarters of a length. Philuminist, third in his previous hunter-chase, got no farther than the first this time, where he fell heavily, pinning Joan beneath him.

When a screen similar to those used when a horse is killed was put up round Joan's prostrate body it was the last straw for her distraught mother. Watching a daughter racing is anxiety enough for an anxious mum, but poor Mrs. McCall thought Joan had been killed.

Joan spent three weeks in traction and two months on crutches but came back determined to enter the fray again.

She chose to ignore the remark of one doctor to another, which she overhead as she lay on a stretcher: "I wouldn't have thought this pelvis *could* be dislocated with thighs like those!"

Unfair, when we note that Joan can scale the minimum 9st 7lb, but it didn't rile her – she endured far less complimentary remarks from the "Duke"!

Joan Barrow did a great job as women's ambassadress in NH racing and the television programme was an unqualified success.

By becoming a household name, Mrs. Barrow increased respect for the sport among people who hadn't a clue to what is really involved.

CHAPTER ELEVEN

VAL GREAVES, ANN HARVEY, GERALDINE REES AND BARBARA OLIVER

VAL GREAVES has a bubbling cheerfulness and love of life, coupled with courage and determination, which make her ideal material for the jumping game. She did much to help the early female image, by displaying ability without a trace of cockiness.

Val is a farmer's wife from Northallerton in Yorkshire and her interest in racing began in the point-to-point field. As a 15-year-old, too young to race, she was schooling after a meeting and rated it kid's stuff. Then, pulling up her irons and adopting a Lester Piggott stance, she fell off at the next open ditch! By the time the Sex Discrimination Act arrived she had had a few point-to-point winners and enjoyed riding out for a nearby permit-holder, Mr. David Barron.

It was he who gave her her first ride under NH rules on Silver Gal, a grey filly. Val had schooled over hurdles and thought she knew what to expect when she became the first woman to ride in a hurdle race.

That was a novice event at Catterick on Valentine's Day 1976 and such a cracking pace was set that Val was left behind and stayed there: "always tailed off", as the form-book unflatteringly described it.

There were 24 runners with professional jockeys like Paddy Broderick, Martin Blackshaw, Denis Atkins, Brian Fletcher, Colin and Nigel Tinkler, David Munro and Jonjo O'Neill in the line-up all experiencing a woman beside them for the first time . . . in a race, that is.

Down at the start, Paddy Broderick was happy to offer advice.

"But," he lilted in his Irish brogue, "don't follow me, I'm going like shit!" He was on the favourite, Mount Blessed, but was beaten on the run-in by 12-1 chance Naivasha. The professionals took little notice of Mrs. Greaves in that race, but they were to do so before long. A few more rides in novice hurdles taught her much more about "riding a race" until in April, with Silver Gal sporting blinkers and Mrs. Greaves much improved, the mare ran really well in a selling handicap at Perth under a featherweight 9st 7lb. Two from home the mare came through to challenge, stood back too far, hit the top . . . and over she tipped.

Val was cursing unashamedly as she picked herself up. A fall "hurts" so much more when the jockey is sure he would have won. Mrs. Greaves was left with an eye which looked as if she had been sparring with Muhammad Ali.

Undaunted, the pair lined up again at Kelso but were brought down at the first, where Major Rex and Stanegate fell. Back to Scotland in early May and Silver Gal again started at 25-1 (the bookmakers had ignored her good Perth run) for the second division of the Yarridge novices hurdle with 11 runners. Rigorous started at odds-on, followed by Major Rex.

There was no hard luck story this time. Val Greaves brought Silver Gal with a steady run from three out, riding for all she was worth and fighting out the finish in splendid style. She forced the mare home by two short heads from Snow River and Rigorous.

The performance brought the praise it deserved. Val Greaves wasn't just the first woman to beat professionals and to win a hurdle race. She had done it by taking them on at their own game and winning on merit. As a result of the inevitable publicity the professionals, with whom Val had been riding and joking for the last few months, learned for the first time that she was a mother of two – an eight-year-old daughter Alex, and Vernon, who was seven. Now, instead of "hey, love!" the other jockeys were at pains to refer to "Mrs. Greaves!"

Val likes to think that if she had been born ten years later she might have given professionals a run for their money. But as it was she was content to go home to her family and their zoo of pets, thankful for husband Ernie's support, instead of poring over form books, wondering when the next ride would turn up.

★ ★ ★

Ann Harvey, daughter of Doncaster trainer Eddie Magner, is a woman who deserved more success. Conceived in Ireland (where her mother won a point-to-point while carrying her) Ann was born and grew up in England, surrounded by racehorses and racing. She was a successful point-to-point rider and jumped at the chance of riding under NH rules the moment the law changed. With her long blonde hair tied back in a pony-tail, she soon showed the male jockeys that she would give no quarter on the racetrack. Ann accumulated 40 rides in the first season and a half, thanks to the encouragement of her husband Harold, her father – and the backing of the stable's patrons.

A comparatively unemotional woman, Ann suffered little from nerves. She simply loved chasing. Even a fall in front of the field on Silberto in a chase at Market Rasen, that relaxed friendly track in Lincolnshire, failed to daunt her. Though kicked along the ground like a human football, she came up smiling with nothing worse than bruises. And she was gratified when two of the jockeys. G. V. Kelly and Glen Graham, went out of their way to ask if she was all right:

"We didn't mean to hit you", they said afterwards.

"No, of course you didn't, and I'm fine, thanks," Ann smiled.

Ann has almost sampled the sweet taste of success on two occasions. When bravely riding Charming Scot at Market Rasen on Boxing Day 1976 (her mount had never run in a chase) she was about ten lengths behind the leaders coming down the final hill and into the tricky last bend three from home. She gave him a sharp slap down the shoulder and the horse responded in fine style, flying the last three and failing, all out, by only two short heads.

This was followed by more places at Leicester and Doncaster but Charming Scot fell in the four-mile NH chase at the Cheltenham Festival meeting (another example of the jinx on women riding there) and again at Ascot.

After this Charming Scot reverted to hurdles to restore his confidence, placing twice more and being ridden, as usual, by Ann. After a race on him at Hexham she was hauled before the stewards, along with Joseph Curtin riding Graffiti. Both had taken the wrong course after the second flight and continued to race. They finished fourth and third and were both disqualified, being informed icily: "In future, make sure you acquaint yourself with the course properly." This was one of those stupid, embarrassing things which can happen, but Ann was pretty careful about walking courses after that.

Her best ride and worst luck was Major Sir Guy Cunard's Glasserton. The Yorkshire trainer, one of the finest point-to-point riders of all time, had confined Glasserton to Ladies' point-to-points, winning all nine in 1975 but missing 1976 through leg trouble. Ann Harvey had the ride on him in the Spring of 1977, winning a point-to-point before taking the mount in a Teesside hunter-chase.

Glasserton showed himself a cut above his rivals, brushing them aside contemptuously and approaching the second last fence still cantering. Ann had a double handful, she was a good four lengths up and the race was, apparently, at her mercy. Glasserton felt equally confident, but this was his undoing. He over-jumped and failed to get his undercarriage down in time. He got up from his fall but in a few hours his back became paralysed. The next day the gallant horse had to be put down.

* * *

Lancashire lassie Geraldine Rees should not have ridden against the professionals. But she beat them.

It happened on Easter Monday, 1977, when the jockey booked to ride Twidale in a novice hurdle at Carlisle could not do the weight. Mrs. Rees held an amateur rider's permit with the vague idea of having a ride sometime. She had no medical record book, so was not entitled to ride against professionals. And she wasn't likely to get one with just two rides in point-to-points behind her and the change to A and B permits imminent.

Neither she nor her father, Capt. James Wilson, who trained Twidale under permit, were aware of this as they set off for Carlisle with open minds whether Geraldine would ride or not. They agreed to attempt to engage a professional when they reached the course but there were 11 Bank Holiday jumping meetings that day and jockeys were in short supply. The Wilsons still hesitated, thumbing through the rule book, uncertain whether Geraldine could ride. They sought out the stewards who gave them the all-clear so Mrs. Rees walked the course, changed and weighed out. Then an official asked her for her medical record book.

"But I haven't got one."

"Well, I'm afraid that means you can't ride."

Stunned, Geraldine told him of the stewards' assurance and a hasty eleventh-hour conference was convened.

The other jockeys had already filed out to the paddock while 21 year-old Geraldine waited anxiously for the decision. At last it came: O.K.

Twidale, a string of noughts in front of his name, was unconsidered in the field of 18 but with an unheard-of girl riding, the other jockeys were concerned that she should not prove a menace to them. They asked what her tactics would be.

"I expect I'll jump off", she replied.

None of the others particularly wanted to make the running so this suited them well. Twidale set off in front with his rivals patiently biding their time behind; some of the jockeys firmly believing that the girl was being run away with. But Twidale was revelling in this unaccustomed freedom. He was going easily within himself – a pleasant change from being anchored behind, which he resented. This sensitive four-year-old had been found for Geraldine by her husband, Henry, but the family had begun to think that the horse was useless. Now their eyes were filled with amazement as they watched from the stands. Turning for home Twidale was still in front.

Geraldine could hardly believe it, either. As she headed for the last, she heard one jockey call to another: "Come on, let's get the girl!"

"I don't think I'll let 'em!" thought Geraldine as she rode with a determined flourish into and over the last, keeping up the gallop to pass the winning post six lengths clear.

Staggered and flushed with excitement, Geraldine rode in to an enthusiastic reception from the holiday crowd. Reins relaxed she patted her horse as congratulations came from all sides, including the other jockeys.

Jonjo O'Neill, who had been on Man of Steel, congratulated her warmly. Another confessed: "You floored us completely. Well done!"

This was a fairytale start to a racing career and this was one which continued in the same vein. Geraldine's next ride on Twidale was at Perth and proved a real fingernail-biter. Now firmly barred from riding in professional races, Geraldine had Twidale in a two-mile amateur handicap for which there were eight runners, three of them lady-ridden. Ailie Nisbet finished fourth on Ukundu but Fiona Storey's horse, Stella's Pet, fell at the second.

Geraldine tried to adopt similar tactics and was soon well clear but this time she did not have things all her own way. She was joined by James Nelson on the favourite, Move Up, at half way and the two had a ding dong battle. Move Up led over the last but Geraldine did not give up easily. Gritting her teeth, she and Twidale fought back like veterans instead of raw novices, prevailing by a neck. Still, this was not all. Geraldine made it three wins from three rides at the end of the season, when at Hereford on June 6 Twidale was the 25 lengths winner of a two-mile handicap hurdle from Honourable Enoch (Mr. Jonathan Cambidge) the 11-10 favourite. In between these races, Twidale ran twice more in professional races without success; it really was a case of "Woman works wonders!"

After such an introduction Geraldine Rees could have been expected to go into racing in a bigger way, but she remained loyal to her first love – eventing. She was a member of the British Junior event team in 1973 and won the big Tidworth three-day event on Pressoss two years later when still Miss Wilson. Princess Anne was sixth on the Queen's Goodwill. Geraldine also completed the courses at Burghley and Badminton, the pinnacle of three-day eventing.

She decided that she preferred the greater degree of horsemanship demanded in the tough sport of eventing. It was a greater challenge.

Says Geraldine: "Eventing has shown that, given a chance, women can show the men where to go. The courses are much bigger and more solid than in racing, and they have to be negotiated alone, with nowhere near the same room for error. You can't take an event fence by the roots and get away with it."

Many would disagree, pointing out that you can't take liberties in a steeplechase either, while hurdling introduces greater speed. And the very fact that you are not alone in a race can lead to big trouble, calling for a cool, clear head in the heat of the moment.

But I must admit that I admire eventing people, not just for the hours of patient dressage practice and their precision riding, but for negotiating those solid obstacles in cold blood. Give me the hurly-burly, racing atmosphere any day!

★ ★ ★

Barbara Oliver grasped the chance to ride under NH rules with both hands. She had been an avid racing fan from her earliest years, an interest which had been kindled by her uncle, trainer Maurice Kellett. Barbara was a keen Pony Club competitor. She and a group of friends had their ponies taken round the circle of hunter trials and one-day events by horse transporter Mr. George Stephenson who was like a "second father" to the girls. It was he who found the opening in NH racing for Barbara shortly after the law changed. By that time the 21-year-old girl from Bishop Auckland, Co. Durham, had ridden a few point-to-point winners. Mr. Stephenson put her in touch with permit-holder Mr. W. Storey who was looking for an amateur. He readily put her up.

Barbara may have lacked racecourse experience but not initiative. She rode out regularly for Bill Murray and then for Arthur Stephenson while she studied for an agricultural degree at Newcastle University. Arthur Stephenson was a hard taskmaster, but fair. He allowed her to school a variety of horses and gave her some Flat race rides, including a winner.

But jumping was the name of the game for Barbara, daughter of plant hire contractor Robert Oliver, who also had a farm. His racing interest was virtually nil, a burning passion for traction engines consuming his every spare minute. Mrs. Eva Oliver, though a non rider, encouraged her daughter's efforts, accompanying her to the races. There Barbara had little time for nerves – once she had her own permit to train in 1976 she was the only person allowed in the racecourse stables with her horses.

Barbara did all the preparatory work, only handing over the leading rein to a helper in the paddock while she hurried off to change, weigh out, saddle up and, finally, don crash cap. She found the men's changing room valets only too willing to help. ("They would certainly have helped dress me, too, if they could!") One member of Barbara's family to give her a hand with the horses was grandfather Jonathan Kellett. A sprightly 88-year-old, he regularly mucked out, groomed and fed for her, especially when she was at University.

Only Nicky Lay and Ann Harvey surpassed her number of rides in the 1976–77 season, so Barbara accumulated the necessary experience. This included a number of falls, some resulting in painful chipped bones. The worst was a first flight cropper at Ayr when 26 runners trampled over her and left her somewhat concussed.

When Barbara had the leg up on Lucky Ambition at Sedgefield on December 1, 1976, Mr. Storey had had the horse for only a few days, so Barbara sought out the previous rider, David Munro.

"Well, he pulls like hell but doesn't get the trip", he told her not very encouragingly. "You'll find you can't cover him up."

That was what happened but the race went her way. She was able to give him a breather half way and before she knew it she was first past the post and quite amazed!

San Palestino at Carlisle on December 16 was a different proposition. He was quite well fancied and Barbara was more nervous because "after all, you never know what's going to go wrong." But nothing went wrong; Barbara had ridden the horse before and she allowed him to make all the running, scoring by 30 lengths in spite of a last flight mistake.

Barbara's biggest success and greatest thrill came in December 1977 riding her own mare, Drumeen, a former unsuccessful point-to-pointer and novice chaser. By Cantab, the mare had needed time, being particularly susceptible to knocks. She was brought down twice, leaving her "half a horse", losing her bloom and making her "soft". Barbara persevered and on December 8 she found herself the 10-1 outsider of three in a novice chase at Ayr. Caldbeck, ridden by R. Lamb, was at odds on followed by Dunquetzal, partnered by that competent amateur John Mackie – but on this occasion he was unseated. Lamb was content to let the female partnership cut out the donkey work up front. Drumeen fenced superbly and when Caldbeck confidently came up to challenge, he found he was being outjumped. Almost level at the last the mare again put in the better leap, keeping the advantage to hold on by a neck at the line under Barbara's vigorous driving, a feat which brought her well-deserved praise.

CHAPTER TWELVE

THE FRENCH FAMILY; DIANA GRISSELL

MRS. Shelagh French is a name to reckon with anywhere in the country, but particularly in the south-east where she has point-to-pointed since 1947. Then, and up to 1965 when ladies' opens were introduced, women were confined to ladies' adjacent races, giving them only six or seven rides a year. Two years later the door was opened to them in hunt races, if the hunt concerned allowed them. Most did, but the Beaufort resisted until the Sex Discrimination Act forced their hand. My hunt – the Eridge – was one of the few not to allow women in the first year – a sore point with me, as I was its only lady point-to-point rider at that time.

In 1974 women were first allowed to ride in certain adjacent races (not maidens, for instance). These were run, like the hunt members races, at a minimum weight of 12st 7lb, so the door was creeping open for women in hunt racing before the SDA.

Before I started point-to-pointing, many people warned me: "Watch out for Mrs. French; don't get on her inside." But Mrs. French has never deliberately crossed me, tried to push me out, pull me off or intimidate me beforehand – more than can be said for one or two others! But usually the atmosphere is one of friendly rivalry.

Shelagh French acquired a well-deserved reputation as one of the best point-to-point riders, particularly as a judge of pace. She is a master at bringing a horse through to win with a perfectly timed run. She has had some crashing falls and bad injuries and her nerves appear to have become worse in recent years. Yet she has gone out half-crippled to win more races and achieved her 100th point-to-point success in 1975.

Mrs. French is one of the few of today's women jockeys who did not grow up with horses. She was born in Beckenham (her father worked in London) and first sat on a beach pony when ten or 11. Like many another little town girl she badgered her parents to give her a pony. Then they moved to Keston, on the outskirts of Bromley, to a house with three acres – just about enough for a pony! During the war Shelagh was evacuated to the north of London and improved her acquaintance with horses. But not until after the war did she first visit a point-to-point: she had never heard of them! She soon decided that this was her game and bought her first point-to-pointer, Rare Commotion, from a local riding school for £60. It also gave her her first win, though the result had to be decided in the stewards' tent; Mrs. French successfully objected to Mrs. Dorothy Evatt.

Shelagh was now married to farmer and keen village cricketer John French and found herself permanently in the countryside, hunting with the West Kent. She produced a string of horses, some of them also raced by John. Though not always immaculately turned out, the French horses were always fit enough to win and always a force to respect.

Bearing three daughters put little curb on Mrs. French's point-to-pointing activities. The elder pair, Sarah and Scarlett, were schooling and galloping by the time they were 12 or so and learning a thing or two from their redoubtable mother. "Redoubtable" is a favourite adjective of point-to-point writers when describing Mrs. French.

Had Mrs. French not obtained a permit the stewards would have received the rough side of her tongue. Luckily, she passed her medical without difficulty. Her eldest daughter, Sarah, was not far behind in receiving her amateur rider's permit. Sarah, a law student, gained her degree from University College, London, soon afterwards, going on to study for the Bar. Both rode in a maiden hunter chase at Lingfield on March 3, 1976 – not an auspicious start. Mrs. French's mount, Guinea Model fell at the 13th (where Nicky Ledger also fell with Jocrow) while Sarah pulled up on Ramrock when well behind. Six days later Mrs. French tried again with Guinea Model but didn't get past the first, so she returned the mare to point-to-pointing - and did little better in that sphere.

Mrs. French never hurries her horses and is quite content to take three years or so to produce the goods. One such was Royal and Ancient, an Armagnac Monarch gelding out of a Pay Up mare for whom Mrs. French paid 700gns at Ascot Sales and brought along quietly at five and six, showing a little promise without being pushed. At seven he won a maiden point-to-point and a ladies' open, qualifying for the South-East novice hunter chase confined to horses in East Anglia and the south east who were maidens before the season.

The race was run at Fontwell on Spring Bank Holiday with four of the ten runners lady-ridden. The buzz among the holiday crowd was that Norfolk's Regal Favour, owned, trained and ridden by Mrs. Ann Holman, was the good thing. Mrs. Holman, another intrepid rider, had arrived in the paddock on crutches to ride her horse at the Melton Hunt Club point-to-point the previous week and one can't imagine *that* happening on the professional racecourse! Regal Favour started a 2-1 favourite.

The country's leading point-to-point rider, Mrs. Josephine Bothway (now Mrs. Gurney Sheppard), daughter of leading point-to-point owner Joe Turner, was on Star Buck. Earlier in the afternoon she had been tailed-off on Pennyman, favourite for the open hunter chase. Sarah French on a tiny Ballymoss chestnut, Hadleigh Mill, made up the female tally.

Mrs. Holman looked like having things all her own way as she set off from flagfall, only Gaycroft briefly heading her. Mrs. French lay quite well out of her ground, as was her custom, but those who looked back, beyond the leaders, could see her threading her way through at the crucial moment. She had made a name for herself as a judge of pace and she produced her horse bang on the scene going into the last. Regal Favour led over this one but Royal and Ancient was in top gear and flying. All he needed was a clean jump. He did not get it, met it wrong and lost crucial momentum.

Undeterred, Mrs. French got down to work to make up the three lengths leeway. She cut down the deficit inch by inch and the pair passed the post locked together. No-one in the crowd knew who had won. But as young Sarah French found a final burst of speed from Hadleigh Mill and took third, they did know that this was the first time that women had filled all three places in a steeplechase. Seconds later another "first" was announced. The judge could not divide the two principals and awarded a dead-heat.

Mother and daughter returned to the fray in the Spring of 1977, but when Guinea Model took her chance again in the Lingfield maiden hunters chase, Mrs. French handed over the reins to Sarah. There were nine runners in the heavy ground and Guinea Model was one of two rank outsiders – not surprising in view of two falls from two runs the previous year. But this time Guinea Model, always prominent and jumping well, ran out a comfortable winner, beating the favourite, Cross Colonist, by 15 lengths.

The French family continued to mix point-to-pointing with hunter chasing and Mrs. French gained a third and fourth with Royal and Ancient before running him in a three-and-a-quarter-mile handicap chase at Uttoxeter in May, carrying 9st 8lb. Behind as usual, she started her run two out. By this time the favourite, All Spirit (Ian Watkinson) looked in command, but Royal and Ancient's challenge prevailed in the last strides to win by a neck. The two leaders came perilously close but Mrs. French survived the objection.

Royal and Ancient turned out again at a Market Rasen evening meeting on May 28. He tackled a three-mile handicap chase while Hadleigh Mill ran in the novice chase. Cumbria, Bar Rock, Fine Fellow and Sage Merlin were among the seven runners, so Royal and Ancient was little considered in spite of his previous win. This time he was the early leader, came again to rejoin the leaders at the 15th and drew away on the flat, beating Bar Rock and Cumbria three-quarters of a length and three lengths.

This evening was to see yet another "first" for women riders. Two races later at 9pm – almost dark enough for the horses to need headlights – Sarah French got Hadleigh Mill, the outsider of five, home by a neck and a length-and-a-half from Mural Crown and Jimmy Miff.

As usual, the Frenchs did off their horses themselves before thinking of their own refreshment or celebrations. Shelagh French always changed in her horsebox: "I wouldn't want to change in an ambulance room," she said, superstitiously.

Early in the 1977–78 season the French family were out in even greater strength. Not only had they decided to keep several of their horses running throughout the winter, instead of putting them away for hunting after October, but middle daughter, Scarlett, a personal secretary, now helped with the horses before going to work in the mornings. Scarlett had already scored in a point-to-point and made her debut under NH rules with an A permit, confining her to amateur hurdles and chases.

In April, 1978, Mrs. French achieved another long ambition when all three daughters were point-to-pointing. Lucy, a catering and hotel management student who showed plenty of drive at home was then 16 – and at the Tickham Hunt meeting on April 8 she followed the family tradition in fine style. Riding Hadleigh Mill, she produced her at the last and sprinted ahead on the run-in. A week later, at the West Kent meeting, the three sisters rode together in the ladies race. Eventually mother and daughters may be seen all four together in an amateur race under NH rules – and that's a record which would surely take a bit of beating!

★ ★ ★

Not a single month passed in 1977 without NH racing taking place somewhere in Great Britain. The 1976–77 season ended at Market Rasen on the evening of June 7 and 1977–78 got under way on July 30, just seven weeks later, at Market Rasen and Newton Abbot.

A horsebox trundled out of its Sussex base heading for the Lincolnshire track at the start of the new season. Mrs. Diana Grissell was at the wheel for the 250-mile journey north and two of the stable's five horses, Simms Pimms and Welsh Buda, were loaded behind. Mr. Gardie Grissell, whose career as a professional trainer began that day, stayed behind to look after the remaining horses at Dallington, a hamlet near the small market town of

Heathfield. He had held a permit the previous season, winning three races with Genovese, a hurdler for whom he paid 525gns at Ascot Sales, and who was another to be reformed by the quiet, relaxed style of a small stable.

A "character" horse, Genovese kicked out at cars for the fun of it, would never lead in work at home and had to be held up until the last minute for a run on the race-course. Diana, Gardie Grissell's wife of a few months, mastered this tactic on her former home track of Huntingdon where, as the first lady rider to win there, the executive greeted her with champagne as she unsaddled. That was on March 8, 1977 and there were six other women riding that day – Nicky Ledger, Shelagh French, Ann Harvey, Diana Thorne, Nicky Lay and Mrs. Jennifer Iliff.

But now, on the first day of the 1977–78 season only Diana Grissell represented the ladies at Market Rasen and not much was expected this time. Of the stable's two, Simms Pimms was thought the straighter. Welsh Buda had point-to-pointed for Mrs. Grissell for some seasons without success, though frequently placing, and she had been third on him in a hunter chase at Folkestone in April 1977. But, at 11, he was long in the tooth for a maiden and appeared a one-paced plodder in staying races.

Simms Pimms ran first and was unplaced. Richard Linley had been booked to ride Welsh Buda in the two-miles-five-furlongs novice chase as Gardie Grissell, anxious to establish himself as a public trainer, did not want prospective patrons to think he only put up an amateur – and his wife at that! At the last moment Richard Linley was claimed for Newton Abbot so Gardie agreed that Diana could ride her horse after all. Welsh Buda was the outsider of the seven and was soon trailing.

Then Diana gave him a sharp reminder on the final circuit and he cruised past the next two horses with surprising ease. Now she followed the advice of jockey friend Ian Watkinson by taking the outside on the far side and final bend to find the better ground. A couple of slaps and a superb jump at the second last, and Mrs. Grissell found herself in the lead and flying! Welsh Buda met the last wrong and fiddled it but Diana stuck to him, quickly picked him up and rallied him skilfully to draw two lengths clear of

Portland (H. Owen) on the run-in! The Flat season was still in full swing, so this feat went almost unsung, but knowing heads nodded in recognition of her performance. What a start for husband Gardie's training career.

Next time out at Plumpton, their local track, Diana proved that this was no fluke. Though only four ran she was still virtually unconsidered, starting at 7-1 for the two-mile novice chase. Odds of 5-4 were laid on Andy Turnell's mount, El Padre.

The distance was short of what had previously been considered Welsh Buda's best. Diana rode him like a veteran, content to lie behind the other three, with Welsh Buda barely out of an exercise canter. Going past the stands and stables to the sharp top corner for the last time, Diana let him take closer order in readiness for the downhill run. But she refused to be hurried. Biding her time, close on the heels of the two leaders, she sailed the downhill fences and water jump. As they rounded the bottom turn into the railway fence, the third last, horse and rider really got the bit between their teeth. Taking off a length behind they landed two lengths to the good.

Now the locals began cheering Diana on, while the punters grew uneasy. No Andy Turnell, Diana Grissell was still one of the most stylish women racing, riding short, with confidence, and using her head. But it *was* Andy Turnell and company to whom she now showed a clean pair of heels. She took the open ditch and last in the lead and actually drew away on the flat from Beach Guard (Nick Holman), Lewesdon Lady (Anthony Webber) and El Padre. Her return to the winner's enclosure was tumultuous.

Diana's successes were not gained on ready-made winners. They were the reward for good background work. This Huntingdonshire farmer's daughter began riding out on Newmarket Heath for Tom Jones when she was 14, and later for Henry Cecil. Her first job was as secretary to Arthur Budgett, where she entered Morston for the Derby before he won in 1973. Then she returned to Newmarket to join Sir Mark Prescott.

Diana and her brother, Gordon Gowlett, rode regularly in East Anglian point-to-points but it was not plain sailing. Diana had to rely on "spare" rides, taking plenty of falls and many places before winning a race. Success came on her 76th point-to-point ride with a horse named Trilife.

Diana first rode under NH rules at Cheltenham – a hunter-chase on Mrs. Mary Lou Stanton's King Jesse. She pulled up before the second last, but the outing earned her an invitation, along with the handful of other women who had ridden there, to the annual Cheltenham dinner for the award to the champion jockey. Diana asked if she could bring her husband and was told: "No, the men can't bring their wives – you asked for equality, now you've got it!"

With two early wins, this young couple seemed all set for a future as a trainer–jockey combination when their plans were rudely interrupted. Shortly after her second win on Welsh Buda Diana was forced to retire from the saddle for a reason unknown to male jockeys. She was pregnant. So, at a stroke, the yard lost half its work-force and the stable jockey! Diana gave birth to a daughter on April 10, 1978.

CHAPTER THIRTEEN

JENNY STAMP AND ROSEMARY WHITE

SHE manoeuvred her wheelchair deftly across the straw. In the far corner she hoisted herself on to the manger by pulleys she had rigged up, settling on the wooden edge, legs dangling limply.

"O.K. Bring him over. Swing his quarters round. Yes, I can reach from there; back a step – that's it."

Jenny Stamp took a mane comb from her pocket and began pulling the horse's tail, oblivious to the danger that she was trapped in a corner should the horse kick out. Subconsciously, all she knew was that life must carry on as near normal as possible. She would not be beaten.

Seven months had passed since that fateful point-to-point in April 1977. Her own horse, Aries Law, had tripped on a spread plate as he went to stride away from a fence, just as if he had trodden upon a loose shoe lace. It had happened with a circuit to go at her local meeting, the South Wold, in Lincolnshire, and Jenny was sitting with a double handful while her rivals already appeared to be off the bit.

Dazed and disappointed, Mrs. Stamp tried to pick herself up. But her legs would not move. At first she did not realise how serious it was. This would dawn on her gradually as she spent the next four months in hospital. Her back was broken and her body paralysed from the waist down. For good. In time, hope for improvement faded. Acceptance followed and then came the courage to carry on . . .

By November she had ridden a horse again – hoisted on to her old Pagan God, embedded in an Australian cowboy-style saddle loaned by a friend. She was learning to drive a pony and trap.

That would enable her to get from A to B independently, but she was determined to enter competitions, too . . .

It is hard to get Jenny Stamp on the phone. Most of the day she is out in the stable yard seeing to the needs of the 12 hunters in at livery: plaiting, grooming and cleaning tack from her wheelchair. A sponsored ride, organised by George and Pearl Pickering in which 100 riders participated, raised £900 locally. This enabled the latest electric wheelchair to be presented to Jenny by Lord Yarborough.

Jenny began her point-to-point career in Sussex, where she took in a few liveries at Coolham. A green mare "buried" her every time it ran but she caught the bug and soon rode her first point-to-point winner. This was Wick, owned by Mrs. Gilbert Smith, a former Master of the South Wold. By then Jenny and her husband Christopher, better known as Crick, worked in Lincolnshire together – first for the Earl of Yarborough and later for Mr. Lawrie Kirkby, Master of the Brocklesby.

She taught their children to ride until, at 18, daughter Barbara took over in point-to-points on Potentate, dual winner of the Melton ladies race and successful in her first hunter chase at Market Rasen in 1977 accounting for Raynham and the former class chaser of Gold Cup calibre, High Ken.

At their next stable, that of Mr. and Mrs. Pinney, Masters of the South Wold, Jenny again taught their daughter, Ann Heath, until she also started point-to-pointing.

Crick and Jenny then set up on their own. The Pinneys gave Jenny her first ride under NH rules, on Fair Gleam a chestnut mare with whom Jenny and Crick had done well in point-to-points. Women had only just been allowed to apply for permits and the first hunter-chase ride went to Joey Newton. But Jenny lost no time in obtaining her permit and had it with her when Fair Gleam was due to run at Nottingham on February 23, 1976. Joey Newton had been booked to ride but had a mount in the Melton Hunt Club cross country ride in Leicestershire that morning.

"Better take your kit with you in case Joey gets hurt," said Mrs. Pinney.

That is just what happened and Jenny had the leg up, giving

the mare a great ride finishing fourth. She was second at Market Rasen next time out, the Quality Fair eight-year-old again proving herself a one-pacer at the finish. A longer distance might suit her best so she was routed for the four-mile NH chase for amateur riders at the Cheltenham Festival. This was run on the opening day so Jenny became the first woman to ride there, setting off with 21 rivals, all ridden by men.

Jenny recalls the race vividly: "It was like a Rugby scrum; they set off like idiots, bumping and boring."

Fair Gleam kept out of trouble, jumping foot-perfectly until she flicked the top of the 11th. The stiff Cheltenham fences have little "give" and down she came. The fall resulted in a broken collar bone for Jenny, so she was not just the first woman rider at Cheltenham but the first under NH rules to suffer a broken bone. This was the third time that Mrs. Stamp had broken a collar bone and however much of an old hand one may be, the injury is still extremely painful. With the aid of quick healing ultrasonic treatment, however, Jenny was riding out a week later and raced again in two weeks. She gained two more places in hunter chases on Fair Gleam, who was sold for 4,000gns at the end of the season. Jenny also rode in a couple of hurdle races on her own horse, Simple Sam.

In the Spring of 1977, Mrs. Stamp was in considerable demand by owners under NH rules and for point-to-pointing. Though never scoring, Jenny kept knocking on the NH door. Her turn would surely have come.

But two rides for Mr. Gordon Adcock at Southwell on April 7 were to be her last under NH rules – pulling up Emigré in a hunters chase and finishing unplaced on Osgodby Coppice in a hurdle (the second horse, Alcock, has no connection with the author, presumably deriving his name from his sire and dam. He is by Alcide out of Bird!).

A few days later Jenny Stamp's career came to its tragic and premature end. Her competent riding will be missed in Lincolnshire and further afield, but she is sure to make a success of her new role – she is that sort of woman and is being helped in no small measure by her understanding husband. The dilemma

Plate 9 It's not all plain sailing . . . Denise Owen parts company with her mount in a point-to-point, the schooling ground for many girl riders.

Plate 10 "One less to worry about!" Fay Geddes looks back from eventual winner Marshalsand to the departure of Jessica Wallace and Dunbrody VI.

Plate 11 (above) The winter sport of NH racing provides little glamour – and plenty of mud. Rosemary White after winning on Horoscope at Stratford-on-Avon.

Plate 12 (right) Gillian Fortescue-Thomas made her name as a professional motor racing driver before embarking on her amateur career under NH rules.

Plate 13 Trainer's wife Nicky Lay in action on Loxley Lad. Injuries suffered in a point-to-point put her out of the saddle for most of 1978.

Plate 14 Mrs. Shelagh French leads her eldest daughter, Sarah, over a point-to-point fence.

Plate 15 Down but not out. Jenny Stamp, paralysed by a point-to point fall, takes the reins again seated next to her husband, Christopher.

Plate 16 Joan "The Big Time" Barrow here riding Wellington Bridge.

facing the couple now is whether or not Crick dare risk continuing his own point-to-point career . . .

★ ★ ★

Rosemary White was four months pregnant when she applied for an amateur rider's permit in order to ride Horoscope, her point-to-point mare, in a hunter-chase at Chepstow. Her doctor had advised that she could carry on as normal, racing included.

But a week before Chepstow, Rosemary was unseated from Horoscope in a point-to-point at Siddington. She was the red hot favourite and should have cantered up.

"I fell off like a sack of potatoes," said Rosemary. She gave it best for that year, 1976, concentrating on motherhood. The mare, once loose, had jumped a hedge off the course and bolted into Cirencester where she "banked" a car, writing it off, and leaving a trail of blood for her pursuers to follow. Incredibly, the tough little chestnut needed only one stitch. No wonder she earned the nick-name "Horrors."

Horoscope was not thought likely to make a racehorse – she stood only 15.2hh and was a hopeless jumper as a youngster. Yet she eventually won the first BMW ladies champion hunter-chase final. The mare was eight years old when her breeder, John Daniell, lent her to Rosemary. Later he sold her the mare for £500 on condition that she was not sold on. Horoscope's early point-to-points were littered with crashing falls but she was a super hunter, becoming a specialist at hedges, ditches and open water in the Berkeley country.

Rosemary eventually had her first ride under rules at Wolverhampton in February 1977. Her baby daughter, Nancy, was now in safe keeping at her lovely Wiltshire home. Here husband John farmed on a sizeable scale in the family concern, devoting his equestrian activities to eventing and cross-country team events (being part of the successful Beaufort team which included Capt. Mark Phillips) and to planning Horoscope's campaign.

Conditions suited mudlarker Horoscope at Wolverhampton, where it was so wet that both the open ditch and water jump were dolled off. This resulted in a cavalry charge of the 18 runners which ended in disaster with a four-horse pile up early on. Rosemary, brought down on Horoscope, was buried somewhere underneath. Once again Horoscope galloped off loose, with frightening consequences. Clearing the single four-foot rail off the course, she galloped down the dual carriageway towards Wolverhampton.

Poor Rosemary, dazed, mud-covered but unhurt, jumped into a car with a policeman to pursue the runaway.

"My God, that's it!" Rosemary shuddered, picturing Horoscope killed amid the heavy traffic.

To her surprise and relief Horoscope was found unharmed – being led round someone's front lawn, nonchalantly gazing at the stream of traffic roaring by a few yards away.

The next outing was in the Beaufort point-to-point, a BMW qualifier. Marshalsland, with his usual West Country pilot Fay Geddes, was odds-on to win for a staggering seventh successive year. The favourite was 20 lengths clear at half way and looked like coasting in but Horoscope, so improved from her young days, would not give up. Rosemary knew that nowadays she could rely on her mount's superb fencing, so she asked for, and got, a prodigious leap at the last, beating the favourite half a length.

Now Horoscope was aimed at the Cheltenham Foxhunters and Rosemary sought the advice of Lord Oaksey, who walked the stiff course with her. Again it was so wet that it had the effect of making the fences several inches higher. After quite a bit of jostling Horoscope met one all wrong, hit it hard, and Rosemary sailed over the mare's short neck into space – yet another female casualty at the premier course.

Down but not out, though beginning to feel she would never get round a NH course, the pair went to Taunton. They were quite well fancied but this flat track did not make enough use of the mare's stamina. She finished a poor third. Rosemary became afraid that she was aiming too high, so decided to have one more try, in the Hunters Improvement Society hunter-chase at Stratford – mainly because Horoscope was so well at home, behaving like a lunatic.

This time conditions were virtually unraceable. But the mare jumped like a stag in the bottomless going, cruising on the bit and conserving her energy for a speedy finish. She galloped away from the favourite, Agilitee, and scored by seven lengths. At Towcester next time she came up against the Thorne's useful five-year-old, Spartan Missile, and other good horses. She was not disgraced, finishing second.

By this time the Whites had their sights set on the BMW final at Chepstow in June and, meanwhile, kept the mare to a programme of hunter-chases, finishing third at the Cheltenham United Hunts meeting before re-appearing at Towcester. Here, once again, the class was so good that in spite of her consistent form Horoscope did not start favourite.

Turning for home and full of running Rosemary was confronted by a wall of four horses ahead just as she wanted to make her forward move. She saw a gap on the inside and went for it.

"I'll put you through the f---ing wing," shouted a leading male amateur rider, pulling his horse across to block her path.

Ignoring his unsavoury threat, Rosemary switched to the gap he had made on his outside when he moved in. Horoscope, luckily, is one of those horses who can be "guided through the eye of a needle" and the other rider had moved so far to the inside that he was virtually galloping on the hurdles track.

Rosemary and Horoscope won well but she got a message via another jockey: "Don't try to take the inside again." A nasty touch of sour grapes and quite unnecessary. No harm was being done.

Next time it was Rosemary who, in her own words, was out-jockeyed. She felt she miscalculated a race at Devon and Exeter which she should have won – but hind-sight winners come more easily when one is still learning! The ability to see one's mistakes is the fastest way to overcome them, and less galling than the criticism of a "grandstand jockey" who has no real knowledge of what is happening.

It was now mid-May. The BMW final was still three weeks away and with the ground firming up rapidly, it was important to stop Horoscope getting too fat and keep her fit without suffering sore shins.

Rosemary rode her in a local nature reserve, hacked her on the Downs and, once or twice, had the use of Toby Balding's Weyhill sand-gallop. Rosemary took her out alone six days before Chepstow, something she seldom did because of the mare's tendency to whip round. Horoscope did just that, as lithe as a cat and so small that Rosemary was left sitting in thin air. She hung on to the reins, determined not to let go, but the mare pulled away and galloped two miles home down the tarmac road.

"Well, that's Chepstow gone," thought Rosemary, limping back despondently and knowing Horrors' terrible reputation when loose. But the mare was unhurt and Rosemary, leg badly torn and bruised, was back in the saddle three days later.

The BMW was a new race, run over three miles, three furlongs, and confined to horses which had qualified at selected ladies point-to-point races all over the country. The final carried £1,300 in prize-money and a corresponding race for men, replacing the similar Players Gold Leaf Championship, was to be run for some £3,400 on the same day.

The ground was so firm by Spring Bank Holiday Monday, June 6, 1977, that John and Rosemary White seriously doubted whether they should run their mudlark at all. Horoscope, twice a winner, had to give 17lb in penalties to most of her rivals, but both horse and rider now had more racecourse experience than the opposition. At Chepstow they found the course was in tip-top condition and a credit to clerk of the course John Hughes, who was also responsible for Liverpool and Lingfield. It resembled a billiard table, but the hard morning rain had made it certain to be slippery.

The runners were Marshalsland, whose habit of "missing out one" in most of his point-to-points made him an uncertain banker over fences; Willow Walk, who had useful point-to-point form in the north with Miss Jane Ramsay in the saddle; Potentate, Barbara Kirkby's good ex-chaser; Mill Straight (Janine Peck) and Hargan (Pip Fisher) both well-known West Country combinations; and the little grey Kent outsider Detling – he and his rider Vicky Warren were having their first race under rules.

Rosemary's greater experience paid off. She took the inside

berth (no arguments this time!) and let Detling make the early running, content to lie midfield. She brought Horoscope through to dispute the running two from home with the favourite, Potentate, losing ground and Marshalsland having clouted his way out of contention.

Going to the last Willow Walk was just in front. This was where Rosemary could rely on Horoscope to rise to the big occasion by putting in a splendid leap. Her timing was perfect, easing ahead on the run to the line to win by a clever half-length with Potentate 25 lengths back in third.

This was a fitting climax to the career of this tiny 12-year-old who had come such a long and eventful way since she was reckoned too small and useless even for point-to-pointing . . .

CHAPTER FOURTEEN

THE GRAND NATIONAL – AND CHARLOTTE BREW

SHE crammed a jockey cap low over her eyes, hair cropped short to resemble a boy's, her slim, maturing body engulfed in a big sweater. Shyly she emerged from the men's – the only – changing room, speaking hoarsely if at all.

She cantered her mount down to the start and, before she knew it, was off for the Grand National, the greatest horse race, the most dangerous steeplechase in the world – her dream come true! Twice she soared over Bechers and finally climbed that long, long run-in, every nerve and sinew straining, to be first past the post . . . and to have her sex identified minutes later. The equine hero was unheard of, his young film star jockey none other than Elizabeth Taylor, alias National Velvet, acting out the dream story of many a starry-eyed, pony-mad young girl.

For years that dream was out of reach, but the Sex Discrimination Act changed it all.

So, on April 2, 1977, there was a girl "for real" in the Grand National. It was a very different story from Enid Bagnold's National Velvet. No disguising true sex this time, but long, fair hair, lovely complexion and tall frame striding out to the paddock, or trying to. For Charlotte Brew was mobbed like a pop star. This was more frightening than the National itself.

The National was the realization of 21-year-old Charlotte's lifelong and unfaltering ambition, a dream so remote as to be sheer fantasy until the SDA. Had her determination not been total she must, surely, have succumbed to the unbelievably savage pressure against her participation. The critics came from all quarters and the worst were those who had least qualifications. Bias and bitchiness engulfed her, but the councillor's daughter from the Essex village of Coggeshall would not be deterred and her steadfastness won the admiration and respect of the silent majority. The critics included some women jockeys, who were unlikely to attempt such a feat, and some male Pressmen who talked through their chauvinistic hats, knowing they would never have to face such a test of nerve. When I read some of their reports, before and after the race, I realised that there are male bitches! The biggest fear voiced was that of injury; next was the chance that she would interfere with other horses through inexperience. These positive criticisms by informed racing people could be answered by Charlotte in only one way: "I'll prove I'm capable."

Here was no rich china doll seeking glamour and fame. Charlotte Brew had spent a year preparing for the Grand National, going to endless trouble to discover all she could about the race in advance and getting herself as fit as was humanly possible. On the day of the race some told her to keep behind and then, having done so, other critics claimed that she never really raced. Poor Charlotte was on a hiding to nothing!

Incredibly, some managed to blame her because seven horses fell at the first. Their jockeys claimed they had been told to "go like hell to make sure you keep ahead of the girl" and were then surprised that their horses could not keep their legs in the ensuing cavalry charge!

This may be an exaggerated tale but it is typical of the feelings at the time.

Charlotte Brew's Grand National bid could be said to have come about by chance. She was the eldest of four children and had always longed to race, though she did not come from a racing family. Her father, Richard Brew, was the Opposition deputy leader of the Greater London Council. But both she and her mother, Judith, hunted and Charlotte had been brought up with ponies and had learnt much from Cherrie Hatton Hall – sneaking out of Benenden School for lessons.

Charlotte's parents were somewhat diffident when she first told them that she intended to ride in the National, but once they had accepted the idea they gave her their full support and backing – and breathed a sigh of relief when she returned safely.

As Charlotte could not point-to-point until she was 18, she gained valuable all-round experience in the Pony Club and by eventing. She found a suitable horse on whom to start point-to-pointing – a big, good-looking chestnut called Barony Fort. Barony Fort had run a bit in Ireland and was well-bred, being by Fortina out of an Arctic Slave mare Sunset Slave (who was full sister to Arctic Sunset).

Charlotte won four point-to-points on him, despite his tendency to jump to the right. He was, perhaps, not the ideal ride for a girl as, apart from hanging, he was very strong. It took two handlers to lead him round the paddock. In 1976 Charlotte was toying with the idea of entering him for the Liverpool Foxhunters; he was a superb jumper. Then the law changed and she realised she could tackle those challenging fences herself. Two other women, one of them Ireland's Ann Ferris, were expected to take part so Charlotte did not expect undue publicity. But when she arrived at Liverpool she was hounded by the Press because the others had been unable to participate. Had she but known it, worse was to follow.

The Foxhunters, run over two-and-three-quarter miles, was a hunter-chase open to horses which had qualified with foxhound packs only. There were nine runners and that grand hunter-chase champion and Foxhunters specialist, Credit Call, started a 5-4 favourite. Barony Fort and two others were the outsiders at 33-1.

A delay in the paddock seemed an eternity to Charlotte, but at last they were under way and her confidence grew with the feel of a safe, experienced horse beneath her. Galloping towards Bechers, she gave him a kick for a perfect take-off and he landed on the far side of that mighty drop, still running. She was in touch with the depleted field – three horses fell at the first and another was unseated at the third. At the third from home Barony Fort jumped his way to the front and for a few moments it seemed that history was in the making. But "Baron" ran wide on the last bend and Credit Call and Creme Brule, two of the best and most experienced hunter-chasers in the country, swept by to fight out a great duel. The Rebel, ridden by seasoned amateur Charles Macmillan chased after them. Even so, the head, eight lengths and eight lengths by which Barony Fort was beaten was no disgrace, a point conveniently forgotten by many a year later.

Charlotte had proved her point by taking a reliable, if somewhat slow, horse round Liverpool and finishing fourth, thanks to superior jumping. She had made history by being the first woman to tackle Aintree, jumping Bechers Brook and the formidable Chair fence. After the race someone said to her: "Do you realise you've qualified for the Grand National?" It was true. Though most entrants must have won a race of a certain value within a stipulated time, any horse which finished in the first four of any chase on the National course qualifies automatically.

Charlotte needed no more incentive and for months before the 1977 Grand National the Press carried stories about her preparation for the great test. It produced a deluge of critics too and Charlotte found her race-riding was deteriorating, though she won an amateur chase at Fakenham on Near and Far. A complex had developed.

"I became frightened of doing something silly. Any mistake got glaring publicity and 'nerves' about it made my riding suffer appallingly. Instead of going out and taking the risks necessary to win I just sat there like a dead duck. It was awful."

One incident which got abundant publicity occurred in a hunter-chase at Plumpton. Mrs. Shelagh French blamed Barony Fort's right-handedness when her mount, Prince Tacitus, fell at the third. This was the only pre-National chase for which Charlotte ran "Baron", confining the rest of his preparation work to the East Anglian point-to-point circuits. She did ride a number of other horses in chases. The Plumpton hunter chase was won by the third woman rider in the race, Mrs. Jenny Hembrow on Guiburn. This competent rider finished second on the 13-year-old at Cheltenham later that spring.

Charlotte eventually found a way of overcoming bad publicity by not reading it! Then if someone said: "Did you read such and such bad report?" she would reply truthfully: "No," and laugh it off. But when she was first thrown before the public eye, her natural curiosity made her read what was written about her, even though it upset her.

Not all the publicity was bad. Many praised her pluck and admired her; the inhabitants of Coggeshall adored her and her horse; but her participation at Aintree had become something of a national debate.

One point she felt was a fair comment – that woman are not the physical equals of men, so she set out to prove her worth. She was no puny stripling and she toned up her 5′ 10″ frame for months in advance, undergoing rigorous training in the local gymnasium with weights and metal contraptions under the expert eye of physical trainer Keith Allison. He watched her gallop her horse so that he could see which muscles were being brought into play and develop them in the gym. Smoking and drinking were luxuries of the past, boy-friends received short shrift, and no time was left for her jewellery and dress designing.

Charlotte pounded up and down, swam, rode work and ran relentlessly, day after day. This produced another problem. She was already big for a jockey and, somehow, had to make the 10st bottom weight for the National (what a gift for her critics if she had gone out overweight!) but her muscles were getting so big and hard that she was putting on weight! Eventually, using a tiny 2lb saddle, she made the weight by about an ounce!

Racing people's feelings towards her remained mixed. Some who did not know Charlotte thought the whole thing was a publicity stunt. When news of a link with the race's sponsors, *The News of the World*, became known it strengthened the thought: "Was Charlotte to become a 'shamateur'?" But, in fairness, to accept a fee for *writing* about her experiences could not be called professionalism. Indeed, John Oaksey combines the two roles with the greatest success and makes his racing reports far more alive as a result. Charlotte's amateur status could only be jeopardised if she were paid for *riding*.

David Nicholson made no secret of his view: "A complete horlicks. Women are not physically strong enough to do themselves justice in such a race, and it certainly doesn't want to be such a moderate pair."

Gillian Fortescue-Thomas, the first ladies NH champion after her successes with Stanhope Street, had no time for Charlotte, believing she had created bad feelings among the professionals who earned their living from racing. She thought Charlotte rode only for glory, but that an average eventer could get round.

An article written in *The Racehorse* of April 15, 1977 – an informed equine journal that lacks the sensationalism of the popular Press – commented, as it was entitled to do, that riders "as inexperienced as Miss Brew taking part in important races were simply courting disaster." The writer went on, not to imply but to state categorically that "though she failed to complete, she is doing very nicely writing a book, and a firm is finding her mounts for the Moscow Olympics, but she has about as much experience of three day eventing as race-riding so her chances of making the team are remote, to say the least."

Christopher Poole, writing in *The Evening Standard* the same month was more kind: "Even Charlotte's most biting critics concede that her brave Aintree effort was full of merit, but some professionals still claim that a rider of such limited experience was a potential threat to other Grand National runners and riders." His criticism was not on grounds of sex but of inexperience in the rough and tumble of the big race.

Muriel Naughton, NH racing's first heroine rider, was full of admiration that Charlotte was having a go on a good safe jumper: "I wouldn't be keen personally as the fences are absolutely massive and it's such a cavalry charge; I think she has tremendous guts."

Muriel could fully sympathise with Charlotte over the publicity, the phone constantly ringing and "cameras being thrust under your face."

Ivor Herbert, *Sunday Express* journalist and author with first-hand experience of NH training wrote in January 1976, *before* the first girls had ridden at all, that he thought it "long odds against even the most muscular, wiry, orange-skinned, thigh-smacking, hard-cursing female completing the full Grand National course for years." How nearly he came to being proved wrong, just a year later!

Charlotte Brew had stuck to her guns. She had a good and safe, if somewhat slow, horse on whom she had completed three parts of the National course already and in whom she had complete confidence. She would finish the course, of that she was utterly convinced, and she would confound her critics. More important, she was going to experience the thrill of competing in the world's greatest race.

As far as the lack of experience went, she felt the National was so much a race unto itself that few jockeys had much experience of that type of riding. Here, she thought, her past eventing ex-experience would be useful. There was nearly as much betting on whether or not Charlotte would get round (odds of 20-1 were quoted) as there was on the favourite to win – Andy Pandy from Fred Rimell's powerful yard (successful the previous year with Rag Trade) – and on Aintree's living legend, Red Rum.

Ian Watkinson stuck by Charlotte through thick and thin, regardless of teasing from his colleagues. He gave her as much help and advice as he could. By the day of the race, Charlotte knew almost every blade of the course. She had walked it so many times and then she walked it twice more on the morning of the race. She had conceived a strategy for the race: she would go to the outside to avoid the possibility of interfering with others though Barony Fort's right-handed jumping had been largely cured by this time, thanks to Ian Watkinson's patient schooling, and to get as good a run as possible, aiming to complete.

"I will complete the course and I won't finish last," Charlotte decreed emphatically at the pre-race Press conference. By this time *The News of the World* had supplied her with two bodyguards to keep the Press at bay and to issue Press statements to save her from harassment.

On the night before the race Charlotte saw one of the most hurtful interviews about her on TV. Three women jockeys were introduced as Charlotte's colleagues and asked for their views on her Grand National bid.

"It's ridiculous and shouldn't be allowed," they chorused like parrots. But the viewer was *not* told that they were *Flat* race jockeys who had never ridden in a chase of any description.

Then three trainers warned Charlotte that they would lynch her personally should she jump across their horses in the race. The fences were rapidly becoming the least formidable of the obstacles confronting her!

Not everyone was like that. When Fred Winter was asked his view that gentleman of NH racing, superb trainer and ex-jockey, said quietly: "If professional jockeys can't keep out of *her* way, then *they* shouldn't be in the race" – and he wished her good luck, adding that some girl had to be first.

And after the race some newspapers took a kinder line: "Charlotte Shows 'Em" (*Sunday Mirror*); "Charlie Girl is Smash Hit" (*Sunday Telegraph*); "Triple Rum – And Let's Raise a Glass to Charlotte" (*News of The World*).

A heap more good wishes (and many good luck charms) reached her on the morning of the race, including two telegrams she particularly cherished. One was from Findon maestro Ryan Price, a fearless hunting man and one of the finest trainers. It read: "Courage and determination will see you through." The other was from Bruce Hobbs who, at 17, was the youngest man ever to win the National, on Battleship in 1938. Charlotte had never met either man.

This gave her the moral boost she needed. "Damn the David Nicholsons of this world," she thought and got on with the job.

She walked the course once more, this time in the company of Richard Pitman and Ian Watkinson. On previous occasions she had never ventured really close to the Chair, the biggest fence on the course with a huge open ditch 6 feet wide in front of a 5′ 2″ fence. Now Ian and Richard teased her, clowning and playing the fool. Ian got into the ditch and pretended to be mountaineering to get out of it.

"Stop it, you're tempting Providence," Charlotte shouted.... it was at the Chair that Sage Merlin fell in the race with Ian Watkinson when going well in second place. Sadly, Sage Merlin was killed early in the 1977–78 season.

The nightmare walk from the weighing room to the paddock, during which the police had to ward off the mobbing, was over. Dispelled now were thoughts of criticism or praise. Charlotte

Brew and Barony Fort were about to make history, but such matters are far from a jockey's mind at such a moment. The ripple of Baron's muscles beneath her, the task to be tackled, the tingling excitement of pre-race nerves were all that filled Charlotte's mind now.

The atmosphere was electric. An elderly earl whispered confidentially to her in the paddock: "If I were you, my dear, I'd go on the outside." As if Charlotte had not formulated her plans by now! People did not realise just how much trouble Charlotte had taken in her preparation.

Barony Fort jigged sideways, his big goose rump swinging out of line in the parade. Soon they were down at the start, not far from the stands. Girths were checked, the roll called and the excited, heaving mass of 42 horses began moving into line. Any second now, and they would be off.

Suddenly a protest group marched out in front of them waving banners. Jockeys muttered curses and turned their horses in anti-climax, their tense bodies unwinding as the police swung into action to disperse the unwelcome visitors. The demonstrators were frog-marched away and the course was clear. Fidgety horses eager to start lined up again; it looked like the off this time.

"Hold it!" a cry rang out, and all eyes were fixed on an object lying in the grass a few yards ahead of them. It was a policeman's helmet! Slowly, one bare-headed bobby walked out, picked up the offending piece of headgear and solemnly replaced it on his crown.

It was one of the funniest, truly British sights Charlotte could remember and it took the last remnants of pressure off her.

Before she knew it they were under way and the first three fences came and went. She could remember little of them, though the seven horse pile-up at the first meant that from then on she had the additional hazard of loose horses all round her. But her surefooted mount did not fail her, rising to the enormous fences – and the occasion – like a veteran.

But Barony Fort was already slipping behind. Charlotte could hardly be blamed for starting steadily after so many pre-race threats, but that one-paced stayer really needed a good start to get into the race properly and to stay there for as long as possible.

With a slow start he had no chance of making a real impression and by half way he was tailed off last.

The progress of Charlotte and Baron was being filmed throughout by a camera operating independently of the rest of the TV crew. It was to record for posterity the first attack by a woman on the world's most famous steeplechase. Some people even complained about that, claiming that Charlotte had spoiled the race because of all the publicity surrounding her. As if she could help it!

The pair survived the first circuit with little difficulty. Approaching Becher's for the second time, three horses were ahead of Barony Fort. Two refused and the other fell. A less honest horse could have been forgiven for calling it a day, but Barony Fort slogged on doggedly, now quite alone. He negotiated the Canal Turn, swinging sharply left-handed after it and approached Valentines more and more slowly. Clearly he was losing impetus, and when he attempted the final open ditch – only four tantalising fences from home – he put his foot in the ditch and stopped.

It was Charlotte's moment of despair. Turning him, she rode at it again but Baron would have none of it and Charlotte was forced to call it a day. The pair had gone farther than all the other non-finishers, but Charlotte rated it a complete failure, her hopes of getting round finally dashed, though the ride Barony Fort had given her would be recalled with pleasure in years to come.

Now Miss Brew was forgotten. Up front more Aintree history had been forged and the crowds were going wild with excitement, almost with disbelief, at having seen the greatest Aintree horse of them all win his third triumphant Grand National.

The unforgettable hero was Red Rum who sailed to victory ahead of Churchtown Boy, accompanied only by two loose horses and the most tumultuous cheers ever heard at Aintree. His story is legendary.

Arkle was the greatest steeplechaser but none has surpassed Red Rum at Aintree: bitter-sweet victor in 1973 over the game, luckless, top-weighted Crisp; winner again in 1974, carrying top weight himself, over L'Escargot; runner-up to L'Escargot in 1975, and pipped on the run-in by Rag Trade in 1976.

Then in 1977 Red Rum returned as a 12-year-old to score for a record-breaking third time – ten years after he made his racing debut on the same course in a five furlongs two-year-old seller in which he dead-heated for first place.

CHAPTER FIFTEEN

THE IRISH SCENE: ANN FERRIS, ROSEMARY STEWART, CHARMIAN HILL AND JESSICA WILLIS

IRELAND . . . soft and beautiful, the people warm and friendly, the hunting superb and racing convivial, the pace of life no faster than the donkeys pulling their loads of peat or churns of milk as they amble along little-used roads. How could such a land be in the forefront of anything? But it was a leader in women's racing.

A full year before that vital change of law in English NH racing, the Irish authorities allowed women to apply for permits. They had always been allowed to compete against men on equal terms in point-to-points, though comparatively few took advantage of it – probably because of the amount of dead weight which had to be carried to make up 12st 7lb and the scarcity of ladies only races, where the weight is usually 11st.

Nowhere can the charm of the Irish be seen more clearly than at the races, be it humble Mallow, the tiny track little more than a point-to-point course in the depths of wild Co. Cork, or at mighty Leopardstown, Ireland's "little Chantilly". Typically, the Irish did not greet the coming of women into steeplechasing with a blaze of publicity. For them it was a natural progression from the point-to-point field. Mr. C. E. O'Sullivan, Keeper of the Match Book for the Irish Turf Club based at The Curragh (Ireland's answer to Newmarket), added that no "obstacle" was put in their way. Women had point-to-pointed for so long that people were conditioned to the idea that they could race over fences.

First came some Flat races for women; then, in 1974, they were allowed to ride in mixed "bumpers" races, the amateur Flat races included on Irish NH cards, so called because of the average amateur's tendency to bump up and down in the saddle instead of hoisting the seat and keeping it clear like the inimitable Lester Piggott.

"Bumpers" are an integral part of the Irish NH scene. They play such an important role in giving initial racecourse experience to future jumpers that there has been a lobby for their introduction in England. Arkle himself began in a "bumper". Horses now may only run if they have never run before four years old. This discourages Flat horses who have no real ability for the jumping game.

Mrs. Charmian Hill was among the women who quickly took advantage of the new situation. She was a doctor's wife from Waterford who, for years, had been a tough and respected habitue of the point-to-point circuits. At 56, Mrs. Hill feared she migh tbe turned down, but her licence was granted after her interview with the stewards. A year later she applied for a jumping licence and this was forthcoming without demur – unlike poor Marie Tinkler in England the following year, 1976. In her first "bumpers" race every male rider wished Mrs. Hill luck as they circled down at the start. She finished third of 23.

When Mrs. Hill had applied for a jumping licence, the stewards merely warned that they would frown upon women taking chance rides. But like the English riders who would follow suit a year later, most stuck to their own horses or to those in their care. Mrs. Hill twice finished second in hurdle races. In her opinion women riders should be permitted an additional 7lb allowance in mixed races to compensate for the amount of dead weight they must carry.

Irish women's successes faded a bit after the first flourish, but they became a part of the scene with little fuss, their acceptance being qualified only by a suspicion that women were probably not strong enough in a tight finish.

The Irish also had problems with changing facilities. It was hard to justify the expense of building for a mere handful (58 women held Qualified Riders Licences in 1977, all of them amateur). Caravans are mostly provided now but at the outset Irish women, like the English, were changing in ambulance rooms mixed up with the male casualties of the day. Ironically, two of the biggest Irish tracks – Leopardstown and Fairyhouse, home of the Irish Grand National – did least to provide women with adequate facilities.

The Rooneys hail from Ulster and are a family which is in the best Irish tradition of hunting and racing. Grandfather point-to-pointed, while father, trainer William Rooney, rode over 400 point-to-point winners, retiring in the 1970's. Two of his four daughters, Ann Ferris and Rosemary Rooney (now Stewart) are Ireland's best known women jockeys. Between them, father and daughters had the Northern Irish point-to-point championship pretty well sewn up for many years. From their early teens the daughters came in for their father's rides whenever he was crocked up! A few years ago it took in all Ireland. Ann won the title in 1976 and Rosemary, with 17 wins, in 1977.

Ten years divide Ann and Rosemary and Rosemary is quick to acknowledge that her elder sister is the one who spent most time helping her riding as a youngster – first in leading rein shows and afterwards escorting her out hunting with the Newry Harriers. Rosemary graduated from the Pony Club to represent Ireland in the Junior show-jumping team in Italy but she soon concentrated her talents on racing, point-to-pointing at 15 (there was no age restriction on women in Ireland).

Apart from growing up in a racing family Rosemary's summer holidays were spent with Findon, Sussex, maestro Capt. Ryan Price. He is a family friend so her passion for racing was intensified.

After leaving school Rosemary travelled south to The Curragh, that attractive expanse of heath which is the principal training centre of the Irish Republic and home of the Irish Sweeps Derby. Its soil is rather like that at Newmarket. There she joined Paddy Prendergast senior and spent two-and-a-half years as assistant trainer, gaining invaluable experience of Flat race riding.

She soon made her mark, winning the first two runnings of the International Ladies Stakes at Leopardstown. She also made successful forays into England winning at Kempton and Nottingham. The skill she had acquired on the Flat helped her when she began riding under NH rules.

In July 1977 Rosemary married George Stewart, who show-jumps for Ireland, farms and runs a coal business. Together they "bring on a few young horses".

Rosemary had followed the pattern set by Ann, who also hunted from the age of six and show-jumped. Ann was only 14 when she rode in her first two point-to-points, winning both on Hill Star. At 15 she notched four more wins between the flags. Ann was 18 when she first won the point-to-point championship, taking advantage of injuries which kept her father on the sidelines.

As most of the point-to-points were mixed races on equal terms between men and women, Ann and Rosemary were already accepted in the amateur ranks when the Irish allowed women to participate under NH rules in 1975. Few remarks were passed, even by the professionals. Though Rosemary felt the professionals did not like their presence she says: "They treated us well. Once we got going and they realised that the few women entering the sport were competent and not likely to interfere, they accepted us."

She added: "After all, we had been schooling the horses at home for the professionals for years. I think we were entitled to ride them."

Ann and Rosemary have two sisters, Eyssen (Ross) and Carol (Adams). While not keen on racing, they take an interest in horses. Eyssen is "head man" at home while Carol enjoys a little hunting. Ann has two teenage children, 15-year-old Gaye, who likes hunting and show-jumping and Billy who, at 14, is not all that keen.

"I wouldn't push it," confesses Ann, acknowledging that it can be far more nerve-racking to watch loved ones racing than to ride oneself.

When Rosemary scored in a "bumpers" on a young horse, Mourneview, at Naas she became the first woman to beat men under rules. The Rooneys bred the Cantab gelding, selling him to Joseph McLoughlin as a three-year-old. He would bring even greater fame to Ann when she won the Ulster Grand National on him – the biggest jumping success by any woman to date. By the end of the 1976–77 season, both sisters had each won 15 races under rules (including the Flat).

Rosemary kicked home a hunter-chase winner at Navan on April 9, 1975, eight months before women in England knew they would be allowed to apply for licences. The horse was Burly Robert, owned by Mr. J. Boggs and trained by Mr. Rooney. He beat the 3-1 favourite, Coxcomical, three-quarters of a length at 7-2. Next came 25-1 shot Devil's Walk, ridden by Ann, who placed in three more hunter-chases and won a few point-to-points with her before joining the small Welsh yard of Mike Bishop.

In 1977, with a Gelligaer hunters' certificate, the gelding won the Horse and Hound final champion hunters chase at Stratford at 50-1! A month after that Irish race Ann won a hunter chase on Mrs. Hugh McCann's Canky.

Rosemary's greatest thrill came in a hunter chase at Punchestown in April 1977 when she scored decisively on the previously unsuccessful Corryvreckan (which in 1978 she qualified for the BMW final at Chepstow). The ground was already soft and the race was run in a thunderstorm but, ignoring the torrential rain, the pair set out to make all the running, putting their 11 rivals at such full stretch that, combined with the conditions, seven ended on the floor.

The Irish NH season continues all through the year, which means that the riding championship runs to December 31. So, during 1975, Ann Ferris kept busy, scoring the first of several hurdle race wins when producing rank outsider Portballintrae to score over four rivals in a two-mile-five-furlongs amateur handicap at Navan.

In the only hurdle on the card at Gowran Park in July that year Ann, on Black Marble, had a desperate struggle with the Tommy Walsh-ridden favourite Prince Rudi, who had to concede two stone. Black Marble got the better of the duel by a short head and survived an objection.

After Rosemary's win with Mourneview in a "bumpers", Ann rode him to two hurdle race victories before going for the Ulster Grand National, run at Downpatrick over three miles on March 3, 1976. He had also won and placed in a few chases and was handily weighted on 10st 4lb with Ann in the saddle. Churchtown Boy was top weight on 12st, started favourite, and fell. He would make his name at Aintree the following year – winning the Topham Trophy on the Thursday and turning out fresh and well on the Saturday to finish runner-up to Red Rum in the Grand National. Well fancied for the big race in 1978, he fell.

Mourneview was quite well fancied for the Ulster race, starting at 5-1 joint third favourite of the 12 runners. They set a good gallop, with Mourneview tucked in mid-division until four from home where he moved up smoothly. Forging ahead two out he continued jumping perfectly and scored by ten lengths from the Frank Berry-ridden Harban (11st 3lb), with Nendrum (9st 10lb) two lengths behind and the second favourite Cottage Queen (9st) another six lengths back.

Ann Ferris (who won a chase in England on Dr. W. Fullerton's Alexangle in 1977) deserved every word of the praise heaped upon her. What a pity that she was unable to tackle the Liverpool Grand National itself the following year because Mourneview suffered an injury.

But he turned out, fit and well, to win a three-mile hurdle at Leopardstown on Boxing Day 1977. He is now owned by Ann.

Though the Irishwomen's racing scene has been dominated by these talented sisters, there are others to show promise, notably 21-year-old Jessica Willis.

Jessica is based in Co. Meath, amid some of Ireland's finest hunting country and an unbeatable schooling ground which produces fearless riders (men and women), as well as horses.

It was Ireland which saw the birth of steeplechasing over two-and-a-quarter centuries ago when, in 1752, Edmund Blake and Cornelius O'Callaghan pitted their hunters against each other across five miles of Duhallow hunting country from Buttevant Church to the spire of St. Leger steeple in the heart of Co. Cork.

History does not record who won, but the country has continued its wonderful racing tradition – with men and women sharing the unsurpassable thrills and spills of the finest equine sport.

CHAPTER SIXTEEN

KING'S RHAPSODY – AND ME

I HAD wanted to race since I was four. It was as simple as that. I grew up in the Spa town of Royal Tunbridge Wells, made fashionable in Regency times, and my family were not in the least horsy. Our only connection with horses were the annual spring visits to the local point-to-points. Picnics were packed, a hamper stowed in the boot with wellies, binoculars, sweaters and the rest of the paraphernalia. When we had queued up and parked, a duster or other suitable "flag" was tied to the car aerial as a landmark for brother Tim, sister Patsy, and myself.

Form was hard to study. Little information could be gleaned from the race-cards, though these were eagerly perused as we munched our way through delicious salads, with hot soup to keep the cold at bay. We picked our way to the paddock through the lines of cars, passed the crowd of bookies shouting the odds and stared, fascinated, at the tic-tac men gesticulating wildly in some strange, uncomprehended code. The name of a horse or our favourite colours influenced us in picking our fancy and we would follow its fortunes while a sixpence or so hung in the balance!

Point-to-pointing still has that amateur spirit, the makeshift tented facilities, the crowds in the beer tents, the joy of meeting old friends, the camaraderie between rivals. Here most jockeys and owners really are in the game for the fun of it. Despite the growth of "professionalism", prize money is pitifully low and honour and glory count for more than anything.

Some amateur riders go on to success under NH rules; some young horses make their names on the racecourse after learning the business point-to-pointing; in their declining years old chasers drop to hunt racing – so called because all point-to-points are run by hunts and, for many, are the biggest fund-raisers of the year. They are the seed-beds where are nurtured those staying chasers which used to be the country's pride, now so sorely needed.

My parents, Rex and Margaret Holland, decided I should not learn to ride until I was seven and I couldn't wait to reach that magical age. I can still remember the excitement with which I donned jodhpurs and cap on that birthday. Weekly lessons began, but I was 13 before I had my first pony. By this time my parents had caught the bug – father hunting while mother owned a couple of point-to-pointers. My determination to race grew stronger. A kid in jodhpurs, I was eyeing the last fence on the old West Kent course at Ightham when an elderly man asked: "Are you going to ride in the ladies' race?" I expect he smiled when I replied: "No, but I will one day." Years later, Rough Scot and Roman Receipt jumped that same fence three times with me – both winners!

As soon as the law changed I arrived hot-foot on the scene, though in a minor sort of way. After an early hunter-chase on my husband's grand old horse Log (remembered, perhaps, for dunking Stan Mellor in a water jump) I acquired a taste for hurdling with Roman Receipt. So Tony and I set about looking for a suitable horse to hurdle. We answered a Horse and Hound advertisement from Warwickshire farmer, David Lowe, who was offering his neat, dark bay King's Rhapsody for sale. We took to the horse immediately, though there were several "ifs" about him. He was a five-year-old who had been fired following a suspicion of tendon trouble. He had run over hurdles without success – though placing a few times – from Stan Mellor's yard. Latterly he had worn blinkers, not a good sign, but he evidently jumped well and looked guaranteed to finish "in the bunch". I did not want a horse who would be likely to get tailed off, misbehave, or draw attention to his lady rider in any other way. Eventually we persuaded my mother to invest £725 in him.

King turned out to be a proper gent, perfect on the busy main roads and kind to look after. We guessed he was probably sick of training, so we seldom galloped him, taking him hunting a time or two instead, and cantering in forestry plantations not remotely like the Berkshire Downs around Lambourn. By January 1977 we decided to run him locally, at Plumpton, to get the feel of him though the ground would be heavy and against him.

Mother, as owner and permit-holder, wanted to see how he went before permitting her daughter to ride but Tony had complete confidence in the horse and pointed out: "If Anne rides, at least we'll really know what he's made of," a reference to some professionals who, quite understandably, take the ride (and fee) on a no-hoper and run round in mid-division before telling the owner what he, the jockey, thinks the owner *wants* to hear, not necessarily what he actually thinks of the horse. Few owners like being told that their horse is either no good or a rogue, and to a professional this is simply part of his job. But when the jockey has also looked after the horse, discovering its character and quirks he (or she) is going to give it the best possible chance in the race and a frank opinion afterwards!

King took his chance in a good handicap in the mud, his blinkers discarded. We decided to let him jump off and enjoy himself if he would. My theory was that he had, probably, always been held up early in a race and pushed out at the end of it, confusing and sickening him. This happens to stacks of novice hurdlers. Of course, few horses can last the distance if rushed off in front and judicious timing is important: a jockey's task is seldom simple!

This time King shared the lead, jumping superbly with his ears pricked, until a mistake half a mile from home when the pack passed him effortlessly. The exercise had told us what we wanted to know: he seemed genuine and happy, certainly not reluctant or trying to duck out.

We decided to wait for better ground and to revert to novice events in the hopes of picking up a place or two, giving me experience and all of us fun. Of course, I dreamed of winning too – what jockey doesn't?

A week or two later King went lame, not very badly and not all the time; he always took a short stride on the road as if hating the hard. It was one of those niggling, frustrating lamenesses and we could not diagnose it, making remedy harder. It was not his tendons and the blacksmith could find no bruises or corns in his foot.

We had him X-rayed locally and the prognosis looked grim. King had suspected ringbone, a bony growth on the pastern just where the horse flexes its foot. It might settle down and cause no more trouble when developed, but it was more likely to worsen, grating the joint with every movement. I worked him in soft fields whenever I could, keeping him semi-fit but fearing he might have to be put down or, at best, be given away. The future for a gelding in his position was bleak.

One doubt persisted. There was often heat in the wall of his foot, a sure sign of bother and very different from ringbone. The answer came in early May. The lameness flared up overnight, crippling poor King completely. Without hesitation we called out Kent's specialist horse vet, Mr. Arnold Crowhurst, who went straight for the foot, prodding his scalpel round a tender spot. The trouble was so deep-seated and old that it had to be poulticed for three days before Mr. Crowhurst could return and release the flow of pus from an area the size of a half-crown. It was Friday, May 6.

Though the season was drawing to a close, I consulted the entries with hope renewed and set my sights on a novice hurdle run over two-and-threequarter miles at Fontwell, hoping for firm ground and a weak race between "tail-enders". This was the Bank Holiday meeting on June 6 when races would be spread over 11 meetings across the country. King was also entered for a seller at Stratford-on-Avon nine days previously. If he could be got ready in such a short time, this could be used as a warm-up, giving us the chance to see the Horse and Hound Cup, the champion hunters chase.

After a week of walking King on the road we had a fortnight left to get him fit. The point-to-pointers had already been turned out to grass so he had to work alone. It meant that he never realised he was actually working; no big string cantering in Indian file on the Downs for him. A week before the race we took him to Camber Sands which stretch for miles at the Sussex seaside resort. I walked and trotted him into the distance by the sea-shore, letting him paddle here and there. We had covered roughly two miles when I turned and headed back in a steady canter, gradually increasing the tempo. It was something of a hit and miss affair and hardly orthodox training, but it gave him a good, refreshing blow.

It was time to four-day declare for the seller. Ben Wise, who trains on the South Downs near Polegate, and for whom I occasionally rode out to help me get fit for racing, had often advocated running in sellers.

"A horse will feel far better if he does well in a poor race than if he is outclassed in something better," he maintained, practising what he preached with considerable success. He placed horses shrewdly and had landed many a wager to prove it.

A selling race is the lowest form of racing life, patronised by the worst horses for the least prize-money. The winner has to be offered for auction and the winning owner gets only ten per cent of the sale money above the advertised selling price, the rest going to the racecourse. If the value is £400 and the horse is sold for £700, the owner gets only ten per cent of the £300 excess (£30), plus his winning share of the £400. He can "buy in" the horse himself, but whether he does that or whether the horse is sold outright, the winning owner can often end up actually losing money. His remedy lies in betting. There is also the provision that any horse in a selling race may be "claimed" no matter where he finishes, usually for two-and-a-half times the advertised selling price, ie, £1,000 for a £400 race.

We were not betting people. What would happen if we won, my parents wanted to know? I assured them there was little likelihood of that.

"It'll be an ideal warm-up for Fontwell," I argued. "It's only two miles and he'll need the race."

The only nagging thought was not knowing just how big a drop in class it would be from novice hurdles. This was a handicap seller, open to horses who could have won any number of races. King had won nothing.

We pored over the rule-book trying to take in the full implications of sellers.

"What will happen if somebody claims him?" my father asked.

"It's not done to do that sort of thing, and I don't suppose anyone would pay a thousand for him", I replied.

"But we're not in the swim of things like professional trainers," Tony ventured.

Finally the parents agreed to have the princely sum of £5 each on the horse "as expenses", just in case all the rest fell by the wayside. His price would be long after an absence of four months, with an unheard of female amateur aboard, and from an obviously non-betting stable.

At four days 21 horses were left, but only ten were declared overnight, just a nice size. Friday, May 27 – just three weeks to the day after King was crippled in his box – dawned clear and sunny. Soon it was blisteringly hot, the first real taste of summer. King left his East Sussex stable early but sweated so much on the journey from the heat that he shed several pounds. During our drive up, Mother glibly promised a slap-up meal at The Bull at Bisham, near the Thames, should we win.

"Well, that's quite safe," I laughed.

"It'll be more like fish and chips if we win. We'll probably end up losing money," Tony added.

I had wondered how to deter people from buying my pal in the unlikely event of victory. We would not want to lose him. I could perhaps dismount to give the impression that he was lame – anything to discourage would-be buyers!

Arthur Davidson's Red Ambion was the only horse weighted above King, who was on 11st 4lb, reduced to 10st 11lb by my allowance. It was soon apparent that the money was down on Knave of Hearts, ridden by Ron Hyett and trained by Martin Tate. He was 3-1 favourite with Kath's Bounty (B. R. Davies) at 7-2. King was generally quoted at 20-1 and was obtainable at 33s, along with two other outsiders.

King looked magnificent, his coat gleaming from the work put into it by our young lad, Peter Guest. His well-being was even more evident. He had shed some weight on the sands, but had obviously lost more in his sweat box on his way up to Stratford. As soon as I was legged up he was jogging and raring to go,

Plate 17 The exceptionally good novice rider, 20-year-old Lucy King from Suffolk, emerged in the 1977/78 season and wound up as the new Ladies NH Champion. With five hunter 'chase winners she beat off the challenge of the two West Country girls, Jackie Thorne (four winners) and Pip Fisher (three winners). Here, Lucy is on Clonmellon.

Plate 18 The first running of the Ladies' BMW Final, at Chepstow in 1977, produced a thrilling finish. Here, at the last fence, are Rosemary White (*left*) on Horoscope – who went on to win – and Jane Ramsay who was second on Willow Walk.

4

Plate 19 Aintree Amazon – Charlotte Brew and Barony Fort clear the water first time round in the 1977 Grand National.

Plate 20 The author pilots King's Rhapsody at Fontwell.

positively pulling me on to the racecourse, quite unusual for him. My parents put on their "expenses" while Tony pushed the boat out with £1 each way, his "luck" money!

King bowled along in front, jumping superbly and going easily within himself. Every time a horse drew near I slapped him down the shoulder and urged him vocally: "Go on, King."

I wanted to keep him in front for as long as possible, afraid he might give up if headed. Two from home I could hear the pack close behind. They would pounce any second now and I just niggled King, clinging to my position for as long as possible. Round the tight final turn and into the last I could hear them pounding closer, but still they did not come by. Half way up the run in, I realised none was going to come! As the form-books would tell it later: King's Rhapsody "made all, ran on well".

Instead of savouring the biggest thrill on earth the worrying thought flashed through my head: "What now? It's a seller!" But the elation would come later, when King was safely home in his own stable. He passed the post three lengths clear of Hello Sailor, Red Ambion and, in fourth, Knave of Hearts. It took me so long to pull him up that there could be no question of pretending he was lame!

The auction was murder. At first there was no bid. I crossed my fingers. I had already run a hand down one leg as ostentatiously as possible, and when kind people, the other jockeys included, congratulated me, I'm afraid I was only able to say: "He's been lame for four months." I was so anxious to stop anyone wanting to buy him!

The auctioneer's voice droned on.

"Only six years old." Then, as an after thought, remembering a woman had ridden him, he added: "He'd make a fine ladies point-to-pointer."

A hand went up. He had procured a bid. Mother countered quickly. 500gns. 550. Not until 725 did the bidding finally stop in her favour.

Then the champagne flowed! The Tote paid 30-1 and with a fiver each way my parents "covered the expenses" and treated us to a memorable meal at Bisham, too!

It did not matter that this was a modest, unimportant race. It was a family victory, the first for all of us, horse included, and it would linger happily in our memories for ever.

Nine days later, King's Rhapsody went to Fontwell a superbly fit and keen horse. Our hopes were quite high. But rounding the top tight turn he was struck into from behind, almost severing a hind tendon. He spent the next four months standing patiently in his box. That's racing. But he lives to fight another day.

His jockey, meanwhile, retired to stud . . . and on January 18, 1978, produced George Warren, who weighed in at 7lb 2½oz, a born winner!

APPENDIX I

STATISTICS

SEASON 1975–76 (FROM JANUARY 30)

145 horses ridden by 41 women; 10 firsts, 15 seconds, 15 thirds, 12 fourths, 14 falls, 3 brought down, 1 failed to start, 3 unseated rider, 75 also ran. Eight different women won. Hunter chases accounted for 7 firsts, 13 seconds and 6 thirds. Big races contested included: Cathcart champion hunters chase, Cheltenham; N.H. Chase, 4m, Cheltenham; Foxhunters, Aintree; Horse and Hound champion hunters chase, Stratford. Champion rider: Mrs. Gillian Fortescue Thomas 3 wins.

SEASON 1976–77

418 horses ridden by 85 women: 42 firsts, 28 seconds, 32 thirds, 28 fourths, 42 pulled up, 30 falls, 2 ran out, 2 brought down, 3 slipped up, 11 unseated rider, 3 refused, 195 also ran. Twenty different women won. 22 hunter chase wins. Big races contested included: Grand National, Aintree; 4m N.H. Chase, Cheltenham Foxhunters, Kim Muir Memorial Chases and Cathcart hunters chase, Cheltenham. Joint champions: Mrs. Nicky Lay, Mrs. Geraldine Rees, and Mrs Rosemary White 3 wins each.

SEASON 1977–78

547 horses ridden by 117 women: 37 firsts, 44 seconds, 40 thirds, 52 fourths, 62 pulled up, 39 falls, 21 unseated rider, 2 slipped up, 2 ran out, 1 carried out, 1 brought down, 245 also ran. 25 different women won. 17 hunter chase wins. Big races contested included: Horse and Hound amateur riders' hurdle, Newton Abbot; Mecca Bookmakers £10,000 handicap hurdle, Sandown; Whitbread Gold Cup, Sandown. Champion rider: Miss Lucy King 5 wins.

APPENDIX II

THE WINNERS

SEASON 1975–76

Feb 7 1976 Nimrod hunters chase, Stratford, 3m2f, Ben Ruler, Miss Diana Thorne, 9-1, 14 ran, neck and 6 lengths from Air General and Miserable Jim.

Feb 17 Air Wedding h.c. Warwick, 2m4f, Indian Diva, Miss Jane Thorne, 6-1, 11 ran, 6l, 2½l from Greystoke Pillar and Menaphon.

Apr 14 Merlin h.c. Ascot, 3m, Stanhope Street, Mrs. Gillian Fortescue-Thomas, 8-11, 9 ran, ½l, 25l from Playbill and Nicky Brown.

May 5 Yarridge nov hurdle, Hexham, 2m, Silver Gal, Mrs. Valerie Greaves, 25-1 11 ran, head, head from Snow River and Rigorous.

May 6 Callow h.c. Hereford, 3m, Stanhope Street, Mrs. Gillian Fortescue-Thomas, 5-6, 8 ran, 8l, 25l from Esoteric and Just Jake.

May 12 West of England h.c., Devon and Exeter, 3m1f, Galloway Fabulous, Miss Katie Halswell, 7-4, 8 ran, 5l, 20l from Cornish Princess and Vulkie.

May 19 Ludlow Gold Challenge h.c., Ludlow, 3m, Stanhope Street, Mrs. Gillian Fortescue-Thomas, 1-2, 8 ran, 5l, 30l from Badsey Brook and Water Sport.

May 31 South East novice h.c. final, Fontwell, 3m2f, Regal Favour, Mrs. Ann Holman, and Royal and Ancient, Mrs. Shelagh French, dead-heat, Regal Favour 2-1, Royal and Ancient 8-1, 10 ran, 15l from Hadleigh Mill.

May 31 Whitsun amateur riders handicap hurdle, Hereford, 2m, Berostina, Mrs. Nicky Lay, 33-1, 12 ran, ½l, 8l from Ten Knots and Delbounty.

SEASON 1976–77

Aug 2 Torbryan selling h'c. c, Newton Abbot, 2m, Jim Hardy, Mrs. Jenny Hembrow, 7-2, 4 ran, 2½l, 20l from Rock Eton and Cwrt Bleddyn, no bid for winner.

Sept 23 Tay h'c. h, Perth, 2m, Deep Mystery, Mrs. Nicky Bullock, 14-1, 9 ran, 1l, ½l from Falcons Boy and Gusty Somers.

Sept 25 Derwentwater h'c. h, Carlisle, Deep Mystery, Mrs. Nicky Bullock, 5-1, ½l, 1½l from Spacer and Caldbeck.

Oct 11 Ayrshire Yeomanry a.r. nov h, Ayr, 2m4f, Jackstones, Mrs. Joan Barrow, 2-1, 13 ran, 30l, 1½l from Spartan Sandal and Tamieshanter.

Nov 12 Celtic a.r. h'c. h, Chepstow, 3m, Jackstones, Mrs. Joan Barrow, 7-1, 8l, 3l from Semi-Colon and Pergusa.

Nov 13 West Norfolk a.r. c, Fakenham, 2m5f, Near and Far, Miss Charlotte Brew, 5-4, 5 ran, 7l, 15l from Straight Fair and Ormonde Tudor.

Dec 1 Grindon s.h. Sedgefield, 2m, Lucky Ambition, Miss Barbara Oliver, 14-1, 6l, 4l from Bonsoir and Chico's Special, bought in 520 gns.

Dec 13 Langbaurgh h'c. h, Teesside, 2m, Lochranza, Miss Yvonne Carr, 10-1, 18 ran, 1½l, 1l from Ice Plant and Solid Silver.

Dec 16 Aspatria s. h, Carlisle, 2m, San Palestino, Miss Barbara Oliver, 2-1, 7 ran, 30l, ¾l from Cunning Trick and Dollie Case, no bid.

Feb 7 Clayton h.c. Plumpton, 3m, Guiburn, Mrs. Jenny Hembrow, 8-1, 13 ran, 8l, 4l from Champers Galore and Brough.

Mar 2 Kent h.c. Lingfield, 3m, Guinea Model, Miss Sarah French, 33-1, 9 ran, 12l, 1l from Cross Colonist and Our Marche.

Mar 5 Halmshaw h.c. Market Rasen, 3m, Potentate, Miss Barbara Kirkby, 7-1, 7 ran, 3l, s.h. from Raynham and High Ken.

Mar 7 Garthorpe maiden h.c. Leicester, 3m, Escallop, Mrs. Anna Garfield, 15-8, 11 ran, 3l, 8l from D'Dyke and King Spirit.

Mar 8 Stevenage a.r. h'c. h, Huntingdon, 2m, Genovese, Mrs. Diana Grissell, 7-2, 19 ran, 8l, neck from Security Council and Topping.

Mar 10 Nigel Thorne memorial h.c. Stratford, 3m2f, Indian Diva, Miss Jane Thorne, 10-1, 10 ran, 2½l, 8l from Armoured Car and Air General.

Mar 14 Ayrshire h.c. Ayr, 3m, Of Course, Miss Ailie Nisbet, 25-1, 10 ran, 3,l 7l from Bar Haze and Monarch's Heir.

Mar 29 Ubique h.c. Sandown, 3m5f, Spartan Missile, Miss Jane Thorne, 5-1, 12 ran, 3l, 20l from Hinterland and True Luck.

Mar 31 Philip Cornes HIS h.c. Stratford, 3m2f, Horoscope, Mrs. Rosemary White, 8-1, 11 ran, 7l, bad from Agilitee and Badsey Brook.

Apr 11 Warnell Fell nov h, Carlisle, 2m, Twidale, Mrs. Geraldine Rees, 25-1, 18 ran, 6l, 5l from Mactavish and Royal Tempest.

Apr 11 Elmhurst h'c. h, Newton Abbot, 3m2f, Current Romance, Mrs. Nicky Lay, 33-1, 12 ran, hd, 2½l from Sea Emperor and Dismasted.

Apr 11 Lutwyche h.c. Hereford, 3m, Sutton Surprise, Miss Pamela Kerby, 14-1, 7 ran, neck, 2l from High Prospect and Devil's Walk.

Apr 11 Waterloo nov h.c. Huntingdon, 3m, Agilitee Mrs. Margaret Wallwin, 7-4, 11 ran, 1½l, 4l from Taffy's Spirit and Cheers.

Apr 16 Hugh Barclay mem nov h.c. Ayr, 3m, Cool Thrust, Miss Gillian Minto, 5-1, 8 ran, ¾,l 20l from Of Course and King Bee.

Apr 20 Teignbridge h.c. Devon and Exeter, 3m1f, Caille, Mrs. Nicky Lay, 9-1, 9 ran, 2l, 4l from Conchita and Devon Spirit.

Apr 21 Chase Meredith mem h.c. Ludlow, 3m, Sutton Surprise, Miss Pamela Kerby, 6-1, 13 ran, nk, 1l from Esoteric and His Last.

Apr 30 Nov h.c. Market Rasen, 3m, Agilitee, Mrs. Margaret Wallwin, 5-6, 11 ran, s.h, 10l from Mr. Muchado and Chief Witness.

May 6 Wedmore h'c. c, Taunton, 3m1f, Alexangle, Mrs. Ann Ferris, 11-2, 12 ran, hd, 12l from Lucky Victory and Highland Brae.

May 6 Deer Park h.c. Towcester, 2m5f, Horoscope, Mrs. Rosemary White, 9-2, 10 ran, 2½l, 4l from Front Seat and Spartan Lace.

May 18 Caputhe a.r. h'c. h, Perth, 2m, Twidale, Mrs. Geraldine Rees, 5-2, 8 ran, nk, 20l from Move Up and Hamilton Lad.

May 19 Allestee h'c. c, Uttoxeter, 3m2f, Royal and Ancient, Mrs. Shelagh French, 14-1, 9 ran, nk, 3l from All Spirit and Corrib Road.

May 21 Lincolnshire a.r. mdn h, Market Rasen, 2m, Chocolate Ripple, Miss Amanda Jemmeson, 15-2, 18 ran, 2l, 5l from Crazy Harvest and Billy Frosty.

May 23 Whissendine h.c. Nottingham, 2m6f, Lothian Brig, Miss Valerie Alder, 7-2, 8 ran, 5l, 12l from Esoteric and Ptarmigan III.

May 25 N. Warwickshire nov h.c. Warwick, 2m4f, Sidewinder, Miss Patricia Caney, 7-2, 14 ran, 8l, 2½l from Brave Money and Helpex.

May 27 Dealers s. h'c. h, Stratford, 2m, King's Rhapsody, Mrs. Anne Alcock, 20-1, 10 ran, 3l, hd from Hello Sailor and Red Ambion, bt in 725 gns.

May 27 Cambrian Soft Drinks point-to-point final, h.c. Haydock, 3m, Bow Bridge, Miss Jane Williamson, 3-1, 6 ran, 4l, 6l from Eastern Day and Hilton Gravelle.

May 28 Thompson h'c. c, Market Rasen, 3m, Royal and Ancient, Mrs. Shelagh French, 15-2, 7 ran, ¾l, 3l from Bar Rock and Cumbria.

May 28 Safari nov c, Market Rasen, 2m5f, Hadleigh Mill, Miss Sarah French, 12-1, 5 ran, nk, 1½l, from Mural Crown and Jimmy Miff.

Jun 6 B.M.W. ladies h.c. final, Chepstow, 3m3f, Horoscope, Mrs. Rosemary White, 9-2, 7 ran, ½l, 25l from Willow Walk and Potentate.

Jun 6 Cullompton nov h.c. Devon and Exeter, 2m3f, Sidewinder, Miss Patricia Caney, 4-5, 10 ran, 20l, 7 l from Porters Precinct and Laudon Hall.

Jun 6 Sporting Chronicle ladies ch h.c. Wetherby, 3m, Lothian Brig, Miss Valerie Alder, 4-5, 4 ran, 8l, dist from Fanakapan and Manicou Bay.

Jun 6 Whitsun h'c. h, Hereford, 2m, Twidale, Mrs. Geraldine Rees, 11-10, 7 ran, 25l, 4l from Honourable Enoch and De Musset.

Jun 6 Madley nov h, Hereford, 2m, Gentle Rose, Mrs. Nicky Lay, 4 ran, 4l, dist from Swift Answer and Nicola Lisa.

SEASON 1977–78

Jul 30 Alford nov c, Market Rasen, 2m5f, Welsh Buda, Mrs. Diana Grissell, 25-1, 7 ran, 2½l, ¾l from Portland and Kutuzov.

Aug 10 Adur a.r. h'c. h, Fontwell, 2m1f, Strong Love, Mrs. Nadine Smith, 20-1, 11 ran, 6l, 2½l from Browns Castle and Shoot the Lights.

Aug 12 Worthing nov c, Plumpton, 2m, Welsh Buda, Mrs. Diana Grissell, 7-1, 4 ran, 4l, nk from Beach Guard and Lewesdon Lady.

Aug 17 Dawlish a.r. h, Devon and Exeter, 2m, Shoot the Lights, Mrs. Rosemary Vicary, 10-11, 3 ran, 2½l, 2½l from Scupper and Mutardis.

Sept 10 Horse and Hound a.r. h'c. h, Newton Abbot, 2m5½f, Cwm Castelle, Miss Jackie Thorne, 11 ran, 11-1, 2½l, 10l from Wilmore and Light Inventory.

Sept 10 Ferfusson Foster s. h'c. h, Sedgefield, Crocume, Mrs. Anne Cousins, 8-1, 15 ran, 3l, 1½l from Frankly Yes and Soldiers Field, bt in 1,350 gns.

Oct 28 Waterloo nov h. div I, Sandown, 2m, Fury Boy, Miss Jane Thorne, 20-1, 21 ran, 1l, ¾l from Indian Sol and Burridge.

Nov 8 Burwash nov h, div II, Folkestone, 2m1f, Clothes Line, Miss Brooke Sanders, 33-1, 14 ran, 4l, 15l from Chichester Bird and Mummy's Star.

Nov 23 Hunters nov h, div II, Worcester, 2m4f, Bujoji, Mrs. Nicky Ledger, 9-2, 9 ran, 1l, 5l from Touch of Spring and Think Big.

Dec 5 Eastgate a.r. nov h, Hexham 2m4f, Royaldieu, Mrs. Sally Williamson, 5-1, 13 ran, 1l, 5l from Arctic Runner and Another Glen.

Dec 12 Marchburn nov c, Ayr, 3m, Drumeen, Miss Barbara Oliver, 10-1, 3 ran, neck from Caldbeck, 2 finished.

Jan 16 Waterloo Road h'c. h, Wolverhampton, 2m4f, Gay Season, Miss Brooke Sanders, 33-1, 11 ran, 1l, 1½l from Rotomar Boy and Golden Murray.

Jan 20 Hampton a.r. h'c. h, Kempton, 2m4f, Needcombe, Miss Jackie Thorne, 6-1, 23 ran, 1½l, 8l from Emperor's Gift and Lampshade.

Mar 23 Somerset h.c. Taunton, 3m1f, Silver Ransome, Mrs. Janine Peck, 5-2, 11 ran, 7l, 20l from Patient and Pellinore.

Mar 23 James Seely memorial h.c. Southwell, 3m, Clonmellon, Miss Lucy King, 9-4, 7 ran, 15l, ½l from Watch Night and High Rebel.

Mar 29 Ubique h.c. Sandown, 3m5f, Clonmellon, Miss Lucy King, 100-30, 10 ran, 2½l, 2½l from Midday Welcome and Sailmaker.

Apr 7 Tom Candy memorial h.c. Devon and Exeter, 3m1f, Hargan, Miss Pip Fisher, 5-1, 11 ran, 4l, 1½l from Kahoutec and Dusky May.

Apr 8 Philip Cornes HIS h.c. Stratford, 3m2f, Agilitee, Mrs. Margaret Wallwin, 5-2, 6 ran, nk, 2½l from Horoscope and Toscason.

Apr 15 Geoffry Lawfield memorial h.c. Huntingdon, 3m, The Coalman, Miss Lucy King, 20-1, 16 ran, 1l, 3l from Hinterland and Sailmaker.

Apr 22 Col. John McKie memorial a.r. h'c. h, Hexham, 2m4f, Beau Brigg, Miss Angela Cross, 14-1, 20 ran, 7l, 4l from Noblero and Gordon's Lad.

May 3 Far South West h.c. Newton Abbot, 3m2f, Hargan, Miss Pip Fisher, 13-8, 14 ran, 12l, 10l from See-o-Duf and Alige.

May 3 Vale of Evesham h.c. Cheltenham, 2m4f, Ptarmigan III, Miss Caroline Saunders, 5-2, 6 ran, hd, 7l from Horoscope, Bushwalks Prince.

May 4 B.B.B. & Hereford Light Horse Breeding Society's h.c. Hereford, 2m4f, Bright Chance, Mrs. Rosemary White, 7-4, 8 ran, 2l, 20l from Royal Air and Toscason.

May 12 Baulking Green h.c. Stratford, 2m6f, Bright Chance, Mrs. Rosemary White, 8-1, 15 ran, 3l, 2l from Spartan Missile and Devil's Walk.

May 13 Sporting Chronicle ladies point-to-point championship final, h.c. Newcastle, Hello Louis, Miss Amanda Jemmeson, 7-2, 11 ran, 2l, s.h. from Lothian Brig and Claverings Cross.

May 20 Lissington a.r. nov c, Market Rasen, 2m5f, Shermoon, Miss Josephine Beswick, 20-1, 16 ran, 10l, 1½l from Royal Roseberry and Napazi.

May 20 Theodore West memorial h'c. c, Market Rasen, 3m, Royal and Ancient, Mrs. Shelagh French, 20-1, 8 ran, s.h., s.h. from Luck and Spacer.

May 22 Whissendine h.c. Nottingham, 2m6f, Lothian Brig, Miss Valerie Alder, 7-1, 14 ran, 1l, 4l from Crofter and Lakeside.

May 24 Ingsdon s. h'c. c, Newton Abbot, 2m, Marcia's Mark, Mrs. Jenny Hembrow, 16-1, 9 ran, 2l, 4l, from Port Dancer and Straight Thorn.

May 24 Bulpin a.r. h, Newton Abbot, 3m2f, Arctic Heir, Miss Jackie Thorne, 5-1, 19 ran, 2½l, 2½l from Holemoor Boy and Peter the Great.

May 25 Totnes open h.c. Newton Abbot, 3m2f, Knockabitoff, Miss Pip Fisher, 5-1, 11 ran, s.h., ½l from Marshalsland and Faberstown.

May 25 Newton Abbot mdn h, div II pt I, Newton Abbot, 2m, Penny Change, Miss Jackie Thorne, 14-1, 11 ran, 1½l, 7l from Port of Verona and Bronze King.

May 27 Horace D. Pain memorial a.r. h'c. h, Cartmel, 2m1f, Deep Mystery, Miss Cathy Houlbrook, 6-4, 9 ran, 6l, ½l from Gentle Rose and Forget It.

May 29 Essandem h.c. Fakenham, 3m, Clonmellon, Miss Lucy King, 7-1, 9 ran, ¾l, 3l from Val d'Amour and Black Outlook.

May 29 J. M. Turner nov h.c. Fakenham, 2m, Mr. Mellors, Miss Lucy King, 7-2, 13 ran, 4l, 6l from Shavington and Diesel Jack.

May 29 Hinchingbrooke h'c. c, Huntingdon, 3m, Royal and Ancient, Mrs. Shelagh French, 8-11, 3 ran, 5l, dist from Esoteric and Talassio.

May 29 B.M.W. ladies championship, h.c. Chepstow, 3m3f, Zanetta, Mrs. Finetta Belcher, 4-6, 8 ran, 12l, 1½l from Port n' Lemon and Galloway Fabulous.

APPENDIX III

RACECOURSE "LEAGUE TABLE"

Number of winners ridden by women:

Newton Abbot	8
Market Rasen	7
Stratford	6
Hereford	6

INDEX